LEED ID&C
Exam Guide & Mock Exam

A Must-Have for the LEED AP ID+C Exam: Study Materials, Sample Questions, Mock Exam, Green Interior Design and Construction, Green Building LEED Certification, and Sustainability (Second Printing of the First Edition)

Gang Chen, LEED AP BD+C

ArchiteG, Inc.
Irvine, California

LEED ID&C Exam Guide & Mock Exam:
A Must-Have for the LEED AP ID+C Exam: Study Materials, Sample Questions, Mock Exam, Green Interior Design and Construction, Green Building LEED Certification, and Sustainability (LEED v3.0)

Second Printing of the First Edition: Issued on 1/26/2012. Incorporated the 2nd edition of the "Green Building and LEED Core Concepts Guide."

Copy Editor: Barry Wenger

ArchiteG, Inc.
http://www.ArchiteG.com

ISBN: 978-0-9843741-0-6
Library of Congress Control Number: 2010923937

PRINTED IN THE UNITED STATES OF AMERICA

Leadership in Energy and Environmental Design (LEED)

LEED CERTIFIED
LEED SILVER
LEED GOLD
LEED PLATINUM

LEED GREEN ASSOCIATE

LEED AP BD+C LEED AP ID+C
LEED AP O+M
LEED AP HOMES LEED AP ND

LEED FELLOW

Dedication

To my parents, Zhuixian and Yugen, my wife Xiaojie, and my daughters Alice, Angela, Amy, and Athena.

Disclaimer

This book provides general information about the LEED AP ID+C Exam and green building LEED Certification. It is sold with the understanding that the publisher and author are not providing legal, accounting, or other professional services. If legal, accounting, and other professional services are required, seek the services of a competent professional firm.

It is not the purpose of this book to reprint the content of all other available texts on the subject. You are urged to read other available texts and tailor them to fit your needs.

Great effort has been made to make this book as complete and accurate as possible; however, nobody is perfect, and there may be typographical or other mistakes. You should use this book as a general guide and not as the ultimate source on this subject.

This book is intended to provide general, entertaining, informative, educational, and enlightening content. Neither the publisher nor the author shall be liable to anyone or any entity for any loss or damages, or alleged loss and damages, caused directly or indirectly by the content of this book.

USGBC and LEED are trademarks of the US Green Building Council. The US Green Building Council is not affiliated with the publication of this book.

Table of Contents

Note: Each credit category is described in a standard format, including Overall Purpose, Mnemonics, Core Concepts, Recognition, Regulation and Incentives, Overall strategies and technologies, and **Specific Technical Information**.

Specific Technical Information for **each credit** includes Purpose, Credit Path, Submittals, Synergies, Possible Strategies and Technologies, Extra Credit (Exemplary Performance), Project Phase, LEED Submittal Phase, Related Code or Standard, and Responsible Party.

A **summary** is added at the end of each credit category, including credit name, related points, extra ID point(s), responsible party, and **a mnemonic**.

The information in this section will not only help you to pass the LEED exam, but also **help you to do real LEED projects**. For example, you can use the summary table in this chapter as a tool to assign tasks for your LEED project team members, and the table can help you to easily determine who is responsible for which prerequisites or credits.

Preface

Starting on December 1, 2011, GBCI began to draw LEED AP ID+C Exam questions from the second edition of *Green Building and LEED Core Concepts Guide*. We have incorporated this latest information in our books. See Appendix 5 for Important Items Covered by the Second Edition of *Green Building and LEED Core Concepts Guide*.

There are two main purposes for this book: to help you pass the LEED AP ID+C (Building Design and Construction) Exam and to assist you with understanding the process of getting a building LEED certified.

The LEED AP ID+C Exam has two parts (or sections):

Part One is EXACTLY the same as the LEED Green Associate Exam. It has 100 multiple choice questions and must be finished within two hours (The total exam time for BOTH parts of the exam is four hours). In this book, "LEED AP ID+C Exam Part One," "LEED AP ID+C Exam Section One," and "LEED Green Associate Exam" are used interchangeably since they are EXACTLY the same.

Part Two is the LEED AP ID+C specialty exam. It focuses on information and knowledge related directly to green interior design and construction (ID+C). It also contains 100 multiple choice questions and must be finished within two hours.

Both sections (Part One and Part Two) of the LEED AP ID+C Exam must be taken back-to-back in the same sitting. The only exception is when a test taker fails one of the two parts, he can retake only the failed section of the exam at a later date.

The raw exam score is converted to a scaled score ranging from 125 to 200. The passing score is 170 or higher. You need to answer <u>about</u> 60 questions correctly for each section to pass. There is an optional 10-minute tutorial for computer testing before the exam and an optional 10-minute exit survey.

The LEED Green Associate Exam is the most important LEED exam for two reasons:

1. You have to pass it in order to get the title of LEED Green Associate.

2. It is also the required <u>Part One</u> (2 hours) of <u>ALL</u> LEED AP+ exams. You have to pass it plus Part Two (2 hours) of the specific LEED AP+ exam of your choice to get any LEED AP+ title unless you have passed the old LEED AP exam before June 30, 2009.

There are a few ways to prepare for the LEED AP ID+C Exam:

1. You can take USGBC courses or workshops. You should take USGBC classes at both the 100 (Awareness) and 200 (LEED Core Concepts and Strategies) level to successfully prepare for Part One of the exam. USGBC classes at 300 level (Green Interior Design and Construction: The

LEED Implementation Process) can be taken to prepare for Part Two of the exam. A one-day course normally costs $450 (as of publication) with an early registration discount, otherwise it is $500. You will also have to wait until the USGBC workshops or courses are offered in a city near you.

OR

2. Take USGBC online courses. Refer to the USGBC or GBCI websites for information. The USGBC online courses are less personal and still expensive.

OR

3. Read related books. Unfortunately, there are few official USGBC books on the LEED AP ID+C Exam, and, as of publication, no exam guides with sample questions. *LEED AP ID+C Exam Guide* is one of the first books covering this subject and will fill in this blank to assist you with passing the exam.

To stay at the forefront of the LEED and green building movement and make my books more valuable to their readers, I sign up for USGBC courses and workshops myself. I review the USGBC and GBCI websites and many other sources to get as much information as possible on LEED. *LEED AP ID+C Exam Guide* is a result of this very comprehensive research. I have done the hard work so that you can save time preparing for the exam by reading this book.

Strategy 101 for the LEED AP ID+C Exam is that you must recognize that you have only a limited amount of time to prepare for the exam. So, you must concentrate your time and effort on the most important content of the LEED AP ID+C Exam. To assist you with achieving this goal, the book is broken into two major sections: (1) the study materials and (2) the sample questions and mock exam.

Chapter One covers LEED Exam Preparation Strategies, Methods, Tips, Suggestions, Mnemonics, and Exam Tactics to Improve Your Exam Performance.

Chapters Two and Three cover general information. I use the question and answer format to try to give you the most comprehensive coverage on the subject of the LEED AP exam. I have given you only the correct answers and information to save you time, i.e., you do not need to waste your time reading and remembering the wrong information. As long as you understand and remember the correct information, you can pass the test, no matter how the USGBC changes the format of the exam.

Chapter Four contains the LEED AP ID+C Exam Technical Review, including Overall Purpose, Mnemonics, Core Concepts, Recognition, Regulation and Incentives, Overall Strategies and Technologies, and **Specific Technical Information**.

Specific Technical Information for **each credit** includes Purpose, Credit Path, Considerations and Benefits**,** Submittals, Synergies, Possible Strategies and Technologies, Extra Credit (Exemplary Performance), Project Phase, LEED Submittal Phase, Related Code or Standard, and Responsible Party.

The final section contains sample questions and a mock exam. These are intended to match the latest real LEED AP ID+C Exam as closely as possible and assist you in becoming familiar with the format of the exam.

Most people already have some knowledge of LEED. I suggest that you use a highlighter when you read this book; you can highlight the content that you are not familiar with when you read the book for the first time. Try covering the answer and then read the question. If you come up with the correct answer before you read the book, you do not need to highlight the question and answer. If you cannot come up with the correct answer before you read the book, then highlight that question. This way, when you do your review later and read the book for the second time, you can focus on the portions that you are not familiar with and save yourself a lot of time. You can repeat this process with different colored highlighters until you are very familiar with the content of this book. Then, you will be ready to take the LEED AP ID+C Exam.

The key to passing the LEED AP ID+C Exam, or any other exam, is to know the scope of the exam, and not to read too many books. Select one or two really good books and focus on them. Actually understand the content and memorize it. For your convenience, I have underlined the fundamental information that I think is very important. You definitely need to memorize all the information that I have underlined. You should try to understand the content first, and then memorize the content of the book by reading it multiple times. This is a much better way than "mechanical" memory without understanding.

There is a part of the LEED AP ID+C Exam that you can control by reading study materials: the section regarding the number of points and credit process for the LEED building rating system. Become very familiar with every major credit category and try to answer all questions related to this part correctly.

There is also a part of the exam that you may not be able to control. You may not have done actual LEED building certification, so there will be some questions that may require you to guess. This is the hardest part of the exam, but these questions should be only a small percentage of the test if you are well prepared. Eliminate the obvious wrong answers and then attempt an educated guess. There is no penalty for guessing. If you have no idea what the correct answer is and cannot eliminate any obvious wrong answer, then do not waste too much time on the question, just pick a guess answer. The key is, try to use the same guess answer for all of the questions that you are completely unsure of. For example, if you choose "d" as the guess answer, then be consistent and use "d" as the guess answer for all the questions that are completely unsure of. That way, you likely have a better chance of guessing more correct answers.

This is not an easy exam, but you should be able to pass it if you prepare well. If you set your goal for a high score and study hard, you will have a better chance of passing. If you set your goal for the minimum passing score of 170, you will probably end up scoring 169 and fail, and you will have to retake the exam again. That will be the last thing you want. Give yourself plenty of time and do not wait until the last minute to begin preparing for the exam. I have met people who have spent 40 hours preparing and passed the exam, but I suggest that you give yourself at least two to three weeks of preparation time. On the night before the exam, look through the questions on the mock exam that you did not answer correctly and remember what the correct answers are. Read this book carefully, prepare well, relax and put yourself in the best physical, mental and psychological state on the day of the exam, and you will pass.

Chapter 1

LEED Exam Preparation Strategies, Methods, Tips, Suggestions, Mnemonics, and Exam Tactics to Improve Your Exam Performance

1. **The nature of LEED exams and exam strategies**

LEED Exams are standardized tests. They should be consistent and legally defensible tests.

The earliest standardized tests were the Imperial Examinations in China that started in 587. In Europe, traditional school exams were oral exams, and the first European written exams were held at Cambridge University, England in 1792.

Most exams test knowledge, skill and aptitude. There are several main categories of exams:

1) Math exams testing your abilities to do various calculations
2) Analytical exams testing skills in separating a whole (intellectual or substantial) into its elemental parts or basic principles
3) Knowledge exams testing your expertise, skills and understanding of a subject
4) Creativity exams testing your skills to generate new ideas or concepts
5) Performance exams such as driving tests or singing competitions

LEED exams test your knowledge of LEED as well as some very basic analytical and calculation skills (on average 1% to 2% of the exam questions require calculations). The LEED AP ID+C Exam is <u>like a history or political science test</u>, and requires the <u>memorization</u> of a lot of LEED information and knowledge.

All LEED exams test candidates' abilities at three hierarchical cognitive levels:

Recognition: ability to recall facts

Application: ability to use familiar principles or procedures to solve a new problem

Analysis: abilities to break the problem down into its parts, to evaluate their interactions or relationship, and to create a solution

A LEED exam has 100 multiple-choice questions. You need to pick <u>one, two, three, or even four correct answers</u> (some questions have five choices), depending on the specific question.

The exam writers usually use errors that people are likely to make to create the incorrect choices (or **distracters**) to confuse exam takers, so you **<u>HAVE to read the question very carefully</u>**, pay special attention to words like <u>may</u>, <u>might</u>, <u>could</u>, etc. Creating effective <u>distracters</u> is a key to creating a good exam. This means that the more confusing a question is, the easier it is for the GBCI to separate candidates with strong LEED knowledge from the ones with weak LEED knowledge.

Since most LEED exam questions have four choices, the **distracters** in a strong LEED exam should be able to attract at least 25% of the weakest candidates to reduce the effectiveness of guessing.

For exam takers, it is to your advantage to guess for questions that you do not know, or are not sure about, because the exam writers expect you to guess. They are trying to mislead you with the questions to make your guessing less effective. If you do not guess and do not answer the questions, you will be at a disadvantage when compared with other candidates. Eliminate the obvious wrong answers and then try an educated guess. It is better to guess with the same letter answer for all the questions that you do not know, unless it is an obvious wrong answer and has been eliminated.

2. LEED exam preparation requires short-term memory

Now that you know the nature of the LEED exam, you should understand that LEED exam preparation requires **Short-Term Memory**. Schedule your time accordingly: in the early stages of your LEED exam preparation, focus on understanding and an **initial** review of the material; in the late stages of your exam preparation, focus on memorizing the material as a **final** review.

3. LEED exam preparation strategies and scheduling

Spend about 60% of your effort on the most important and fundamental LEED material, about 30% of your effort on sample exams, and the remaining 10% on improving your weakest areas, i.e., reading and reviewing the questions that you answered incorrectly, reinforcing the portions that you have a hard time memorizing, etc.

Do NOT spend too much time looking for obscure LEED information because the GBCI will HAVE to test you on the most **common** LEED information. At least 80% to 90% of the LEED exam content will have to be the most common, important and fundamental LEED information. The exam writers can word their questions to be tricky or confusing, but they have to limit themselves to the important content; otherwise, their tests will NOT be legally defensible. At most, 10% of the test content can be obscure information. You only need to answer about 60% of all the questions correctly. So, if you master the common LEED knowledge (applicable to 90% of the questions) and use the guess technique for the remaining 10% of the questions on the obscure LEED content, you will do well and pass the exam.

On the other hand, if you focus on the obscure LEED knowledge, you may answer the entire 10% obscure portion of the exam correctly, but only answer half of the remaining 90% of the common LEED knowledge questions correctly, and you will fail the exam. That is why we have seen many smart people who can answer very difficult LEED questions correctly because they are able to look them up and do quality research. However, they often end up failing the LEED exam because they cannot memorize the common LEED knowledge needed on the day of the exam. The LEED exam is NOT an open-book exam, and you cannot look up information during the exam.

The **process of memorization** is like **filling a cup with a hole at the bottom**: You need to fill it faster than the water leaks out at the bottom, and you need to constantly fill it; otherwise, it will quickly be empty.

Once you memorize something, your brain has already started the process of forgetting it. It is natural. That is how we have enough space left in our brain to remember the really important things.

It is tough to fight against your brain's natural tendency to forget things. Acknowledging this truth and the fact that you cannot memorize everything you read, you need to focus your limited time, energy and brain power on the most important issues.

The biggest danger for most people is that they memorize the information in the early stages of their

LEED exam preparation, but forget it before or on the day of the exam. The danger is that they still THINK they remember the information.

Most people fail the exam NOT because they cannot answer the few "advanced" questions on the exam, but because they have read the information but canNOT recall it on the day of the exam. They spend too much time preparing for the exam, drag the preparation process on too long, seek too much information, go to too many websites, do too many practice questions and too many mock exams (one or two sets of mock exams can be good for you), and **spread themselves too thin**. They end up **missing out on the most important information** of the LEED exam, and they fail.

The LEED Exam Guide Series along with the tips and methodology in each of the books will help you MEMORIZE the most important aspects of the test to pass the exam ON THE FIRST TRY.

So, if you have a lot of time to prepare for the LEED exam, you should plan your effort accordingly. You want your LEED knowledge to peak at the time of the exam, not before or after.

For example, if you have two months to prepare for the exam, you may want to spend the first month focused on reading and understanding all of the study materials you can find as your **initial** review. Also during this first month, you can start memorizing after you understand the materials as long as you know you HAVE to review the materials again later to retain them. If you have memorized something once, it is easier to memorize it again later.

Next, you can spend two weeks focused on memorizing the material. You need to review the material at least three times. You can then spend one week on mock exams. The last week before the exam, focus on retaining your knowledge and reinforcing your weakest areas. Read the mistakes that you have made and think about how to avoid them during the real exam. Set aside a mock exam that you have not taken and take it seven days before test day. This will alert you to your weaknesses and provide direction for the remainder of your studies.

If you have one week to prepare for the exam, you can spend two days reading and understanding the study material, two days repeating and memorizing the material, two days on mock exams, and one day retaining the knowledge and enforcing your weakest areas.

The last one to two weeks before the LEED exam is absolutely critical. You need to have the “do or die” mentality and be ready to study hard to pass the exam on your first try. That is how some people are able to pass the LEED exam with only one week of preparation.

4. Timing of review: the 3016 rule; memorization methods, tips, suggestions, and mnemonics

Another important strategy is to review the material in a timely manner. Some people say that the best time to review material is between 30 minutes and 16 hours (the **3016** rule) after you read it for the first time. So, if you review the material right after you read it for the first time, the review may not be helpful.

I have personally found this method extremely beneficial. The best way for me to memorize study materials is to review what I learn during the day again in the evening. This, of course, happens to fall within the timing range mentioned above.

Now that you know the **3016** rule, you may want to schedule your review accordingly. For example, you may want to read new study materials in the morning and afternoon, then after dinner do an initial review of what you learned during the day.

OR

If you are working full time, you can read new study materials in the evening or at night, and then get up early the next morning to spend one or two hours on an initial review of what you learned the night before.

The initial review and memorization will make your final review and memorization much easier.

Mnemonics is a very good way for you to memorize facts and data that are otherwise very hard to memorize. It is often arbitrary or illogical but it works.

A good mnemonic can help you remember something for a long time or even a lifetime after reading it just once. Without the mnemonics, you may read the same thing many times and still not be able to memorize it.

There are a few common mnemonics:

1) **Visual** Mnemonics: Link what you want to memorize to a visual image.
2) **Spatial** Mnemonics: Link what you want to memorize to a space, and the order of things in it.
3) **Group** Mnemonics**:** Break up a difficult piece into several smaller and more manageable groups or sets, and memorize the sets and their order. One example is the grouping of the 10 digit phone number into three groups in the U.S. This makes the number much easier to memorize.
4) **Architectural** Mnemonics: A combination of Visual Mnemonics and Spatial Mnemonics and Group Mnemonics.

   Imagine you are walking through a building several times, along the same path. You should be able to remember the order of each room you walk through. Then, if you want to remember some type of information, you could break up the information into several representative images and imagine you hanging these images on the walls of the various rooms you walk through as you move through the building. You should be able to easily recall each group in an orderly manner by imagining you are walking through the building again on the same path, and looking at the images hanging on walls of each room. When you look at the images on the wall, you can easily recall the related information.

   You can use your home, office, or another building that you are familiar with to build an Architectural Mnemonics to help you organize the things you need to memorize.
5) **Association** Mnemonics: You can associate what you want to memorize with a sentence, a similarly pronounced word, or a place you are familiar with, etc.
6) **Emotion** Mnemonics: Use emotion to fix an image in your memory.
7) **First Letter** Mnemonics: You can use the first letter of what you want to memorize to construct a sentence or acronym. For example, **"Roy G. Biv"** can be use to memorize the order of the 7 colors of the rainbow, it is composed of the first letter of each primary color.

For example, fixtures for Water Efficiency (WE) in LEED system, you can use **Association** Mnemonics and memorize them as all the plumbing fixtures for a typical home, PLUS Urinal.

OR

You can use "Water S K U L" (**First Letter** Mnemonics selected from website below) to memorize them:

Water Closet
Shower
Kitchen Sink
Urinal
Lavatory

Here is another example of **First Letter** Mnemonics:

Materials and Resources (MR) - Mandatory materials to be collected for recycling:

People Can Make Green Promises

Paper
Cardboard
Metal
Glass
Plastics

5. The importance of good and effective study methods

There is a saying: "Give a man a fish and you feed him for a day. Teach a man to fish and you feed him for a lifetime." I think there is some truth to this. Similarly, it is better to teach someone HOW to study than just to give him good study materials. In this book, I give you good study materials to save you time, but more importantly, I want to teach you effective study methods so that you can not only study and pass LEED exams, but also so that you will benefit throughout the rest of your life for anything else you need to study or achieve. For example, I give you samples of mnemonics, but I also teach you the more important thing: HOW to make mnemonics.

Often in the same class, all the students study almost the SAME materials, but there are some students that always manage to stay at the top of the class and get good grades on exams. Why? One very important factor is they have good study methods.

Hard work is important, but it needs to be combined with effective study methods. I think people need to work hard AND work SMART to be successful at their work, career, or anything else they are pursuing.

6. The importance of repetition: read this book at least three times

Repetition is one of the most important tips for learning. That is why I have listed it under a separate title. For example, you should treat this book as the core study materials for your LEED exam and you need to read this book at least three times to get all of its benefits:

1) The first time you read it, it is new information. You should focus on understanding and digesting the materials, and also do an initial review with the **3016** rule.
2) The second time you read it, focus on reading the parts I have already highlighted AND you have highlighted (the important parts and the weakest parts for you).
3) The third time, focus on memorizing the information.

Remember the analogy of the memorization process as **filling a cup with a hole on the bottom**?

Do NOT stop reading this book until you pass the real exam.

7. When should you start to do sample questions and mock exams?
After reading the study materials in this book at least three times, you can start to do mock exams.

8. How much time do you need for LEED exam preparation?
Do not give yourself too much time to prepare for the LEED exam. Two months is probably the maximum time you should allow for preparing for the LEED exam.

Do not give yourself too little time to prepare for LEED exam. You want to force yourself to focus on the LEED exam but you do NOT want to give yourself too little time and fail the exam. One week is probably the minimum time you should allow for preparing for the LEED exam.

9. The importance of a routine
A routine is very important for studying. You should try to set up a routine that works for you. First, look at how much time you have to prepare for the LEED exam, and then adjust your current routine to include LEED exam preparation. Once you set up the routine, stick with it.

For example, you can spend from 8:00 a.m. to 12:00 noon, and 1:00 p.m. to 5:00 p.m. studying new LEED exam study materials, and from 7:00 p.m. to 10:00 p.m. performing an initial review of what you learned during the daytime. Then, switch your study content to mock exams, memorization and retention when it gets close to the exam date. This way, you have 11 hours for LEED exam preparation everyday. You can probably pass the LEED exam in one week with this method. Just keep repeating it as a way to <u>retain</u> the LEED knowledge.

OR
You can spend from 7:00 p.m. to 10:00 p.m. studying new LEED exam study materials, and from 6:00 a.m. to 7:00 a.m. performing an initial review of what you learned the evening before. This way, you have four hours for LEED exam preparation everyday. You can probably pass LEED in five weeks with this LEED preparation schedule.

A routine can help you to memorize important information because it makes it easier for you to concentrate and work with your body clock.

Do NOT become panicked and change your routine as the exam date gets closer. It will not help to change your routine and pull all-nighters right before the exam. In fact, if you pull an all-nighter the night before the exam, you may do much worse than you would have done if you had kept your routine.

All-nighters or staying up late are not effective. For example, if you break your routine and stay up one-hour late, you will feel tired the next day. You may even have to sleep a few more hours the next day, adversely affecting your study regimen.

10. The importance of short, frequent breaks and physical exercise
Short, frequent breaks and physical exercise are VERY important, especially when you are spending a lot of time studying. They help relax your body and mind, making it much easier to concentrate when you study. They make you more efficient.

Take a five-minute break, such as a walk, at least once every one to two hours. Do at least 30 minutes of physical exercise everyday.

If you feel tired and cannot concentrate, stop, go outside, and take a five-minute walk. You will feel much better when you come back.

You need your body and brain to work well to be effective with your studying. Take good care of them. You need them to be well-maintained and in excellent condition. You need to be able to count on them when you need them.

If you do not feel like studying, maybe you can start a little bit on your studies. Just casually read a few pages. Very soon, your body and mind will warm up and you will get into study mode.

Find a room where you will NOT be disturbed when you study. A good study environment is essential for concentration.

11. A strong vision and a clear goal

You need to have a strong vision and a clear goal: to master the LEED system and become very familiar with the LEED certification process. This is your number one priority. You need to master the LEED information BEFORE you attempt any sample questions or mock exams. It will make the process much easier. Otherwise, there is nothing in your brain to be tested. Everything we discuss in this chapter is to achieve this goal.

As I have mentioned on many occasions, and I say it one more time here because it is so important:

It is how much LEED information you can understand, digest, memorize, and firmly retain that matters, not how many books you read or how many sample tests you have taken. The books and sample tests will NOT help you if you cannot understand, digest, memorize, and retain the important information for the LEED exam.

Cherish your limited time and effort and focus on the most important information.

Chapter 2
Overview

1. **What is LEED? What is the difference between LEED, LEED AP, LEED Green Associate, LEED AP+ and LEED Fellow?**
   Answer: LEED is a term for buildings. It stands for the Leadership in Energy and Environmental Design (LEED) Green Building Rating System™. It is a voluntary system set up by the U.S. Green Building Council (USGBC) to measure the sustainability and performance of a building.

   LEED AP, LEED Green Associate, LEED AP+, and LEED Fellow are terms for people.

   LEED AP stands for LEED Accredited Professional. A LEED AP is a person who has passed at least one of the three old versions of the LEED exam (LEED-NC, LEED-CI, and LEED-EB) before June 30, 2009. This person has the skills and knowledge to encourage and support integrated design, to take part in the design process, and to control the application and certification process for a LEED building.

   LEED APs have **three choices:**

   1) Do nothing and keep the title of LEED AP. It is also called LEED AP without specialty, or a legacy LEED AP.

   2) Starting June 2009, a LEED AP can choose to enroll in the new tiered system, accept the GBCI disciplinary policy, and **complete** the prescriptive Credentialing Maintenance Program (**CMP**) for the initial two-year reporting period. After opting in, a LEED AP can use one of the new specialty designations (Building Design and Construction ("BD+C"), Interior Design and Construction ("ID+C"), or Building Operation and Maintenance ("O+M")) after his name. The LEED AP must opt in before summer of 2011.

   3) Starting June 2009, a LEED AP can choose to opt in, accept the GBCI disciplinary policy, **agree** to CMP, and **pass** Part Two of one of the LEED AP+ specialty exams to become a LEED AP+. The LEED AP only needs to take Part Two of the LEED AP+ exam if he takes the exam by summer, 2011. After opting in, he can use one of the new specialty designations (BD+C, ID+C, or O+M) after his name.

   By choosing Paths 2 or 3 above, a LEED AP will become a LEED AP+ or a LEED AP with Specialty. See detailed information below:

   A LEED AP who passed the old LEED-NC exam will become a LEED AP BD+C, or a LEED AP BD&C.

   A LEED AP who passed the old LEED-CI exam will become a LEED AP ID+C or a LEED AP

ID&C.

A LEED AP who passed the old LEED-EB exam will become a LEED AP O+M or a LEED AP O&M.

Maintenance: After completing one of these three choices, a LEED AP will be treated like any other LEED AP+ and will need to pay a $50 fee and take 30 hours of required class every two years to maintain the title of LEED AP+. The $50 fee is waived for the first two years for a Legacy LEED AP who decides to opt in.

LEED Green Associate, LEED AP+, and LEED Fellow are the three new tiers of professional credentials for LEED professionals. The GBCI started to use these new designations in 2009.

A **LEED Green Associate** is a green building professional with a basic level of LEED knowledge, i.e., a person who has passed the LEED Green Associate Exam and possesses the skills and knowledge to understand and support LEED projects and green building in the areas of design, construction, operation, and maintenance, AND has signed the paperwork to accept the GBCI disciplinary policy.

The LEED Green Associate Exam will NOT test the detailed information for each LEED credit, i.e., WE2.1, WE2.2, etc. It will test you on the overall core concepts. You need to know the strategies for the overall categories of WE, EA, etc.

However, it IS much easier to understand, digest and organize LEED information, and it will be to your advantage to learn and memorize the LEED information for each specific LEED credit. You will be a much more desirable and useful support member for a LEED project team. You will definitely be able to answer generic LEED questions for the major LEED categories and pass the LEED Green Associate Exam (LEED GAE).

You also need to know the codes and regulations related to each of the main LEED categories, i.e., WE or EA, etc., but NOT information for each credit within the category.

Exam Cost: The cost of LEED Green Associate Exam is a $50 non-refundable application fee plus $150 for USGBC members or full-time students and $200 for non-members.

Maintenance: A LEED Green Associate will need to pay a $50 fee and take 15 hours of required class every two years to maintain the title.

LEED AP+ is a green building professional with an advanced level of LEED knowledge, i.e., a person who has passed the LEED Green Associate Exam (or Part One of the LEED AP+ exams) and Part Two of a LEED AP+ specialty exam based on one of the LEED rating systems or the equivalent, AND has signed paperwork to accept the GBCI disciplinary policy.

There are five different categories of LEED AP+ specialty exams (Part Two of LEED AP+ exams) and five categories of related LEED AP+ credentials, including:

LEED AP BD+C (Building Design and Construction)
LEED AP ID+C (Interior Design and Construction)
LEED AP O+M (Operation and Maintenance)

LEED AP ND (Neighborhood Development)
LEED AP Homes

For example, if you want to become a LEED AP BD+C, you need to pass the LEED Green Associate Exam (Part One of ALL LEED AP+ Exams) and the LEED AP ID+C Specialty Exam (Part Two of LEED AP+ exam specializing in LEED BD+C).

Both Part One and Part Two of the LEED AP+ exam have 100 multiple-choice questions, asking for one, two, three or even four correct answers (Some questions have five choices).

Exam Cost: The cost of the LEED AP+ exam is a $100 non-refundable application fee plus:

1) If you are taking the combined exam (both Part One and Part Two): $300 for USGBC members and $450 for non-members.
2) If you are taking only the Part Two specialty exam: $150 for USGBC members and $250 for non-members.

Maintenance: A LEED AP+ will need to pay a $50 fee and take 30 hours of required class every two years to maintain the title.

A **LEED Fellow** is a green building professional with an extraordinary level of LEED knowledge, and has made major contributions to green building industry. The GBCI is still developing the criteria for becoming a LEED Fellow.

Note: These are current fees. They may be changed by the GBCI at a later point, please check the GBCI website for exact fees.

2. **Why did the GBCI create the new three-tier LEED credential system?**
Answer: To pursue on-going improvement and excellence, to assure stakeholders of LEED professionals' current competence and latest knowledge in green building practice, and meet three prevailing market challenges: stay current, differentiation, and specialization.

3. **Do I need to have LEED project experience to take the LEED exams?**
Answer: If you are taking LEED Green Associate Exam, you need to meet one of the three criteria below:

   1) Studying in an education program that addresses green building principles

   Per USGBC:
   "GBCI has confirmed that all of USGBC's education programs, with the exception of its webinars, satisfy this requirement." Refer to the link below for more information:
   http://www.usgbc.org/DisplayPage.aspx?CMSPageID=2011

   This means ANYONE can take the LEED Green Associate Exam.

   Why?
   Because ANYONE can take the one of the USGBC's education programs listed at link above and qualify for the exam as long as he is willing to pay the fee (about a few hundred dollars, the exact fee depends on which program he chooses). A candidate probably will take one of these education programs to prepare for the LEED exam anyway.

OR
2) Working in a sustainable field

OR
3) Have previous experience supporting a LEED-registered project

This experience must be confirmed by a dated letter on letterhead stationary, in 750 words or less for the LEED Green Associate Exam, and 1,500 words or less for the LEED AP+ Exam. The letter must be signed by a client, supervisor, project manager, teacher, or someone else who is qualified to evaluate your performance. The letter's author needs to use a business card or other means to demonstrate his title and relationship to the applicant.

The letter must also provide the date of the applicant's related experience, as well as confirm and summarize the applicant's participation or experience in one of the three qualifying categories (green building education, a LEED project, or sustainable work).

For a candidate qualifying through green building education, an official transcript or a certificate of completion can substitute for a letter of attestation.

5% to 7% of the candidates will be audited for the above requirements.

You also need to agree to the GBCI credential maintenance requirements and the disciplinary policy. See LEED Green Associate Exam candidate handbook for detailed requirements.

If you are taking the LEED AP+ exam, you must have documented experience on one or more LEED projects within three years of your exam application submittal date. See LEED AP+ exam candidate handbook for detailed requirements.

4. **How do I become a LEED AP+? Do I have to take the LEED Green Associate Exam first to become a LEED AP+?**
Answer: There are four paths to become a LEED AP+. See below. Paths One and Two do NOT require taking the LEED Green Associate Exam first, while Paths Three and Four do.

Paths One and Two: If you are one of those lucky people who passed one of the three old LEED AP exams before June 30, 2009, and decided to opt into the new tiered system, you just need to opt in and accept the GBCI disciplinary policy and CMP requirements,

AND
1) **pass** the Part Two of the LEED AP+ exam before summer of 2011 and **agree** to the CMP

OR
2) **complete** the CMP before summer of 2011.

Path Three: If you have already passed the LEED Green Associate Exam, you only need to pass one of the specialty exams (Part Two of the LEED AP+ exams) to earn the title of one of the LEED AP with specialty.

Path Four: If you have not taken any LEED exams, you need take both parts of the LEED AP+ exam. You need to take both parts at the same time (back to back in the same sitting), UNLESS you have passed one of sections before. In this case, you can schedule to retake the only failed section at a different time for an additional fee.

5. **How many questions do you need to answer correctly to pass the LEED exams?**
Answer: Many readers have asked me this question before. The short answer is about 60 correct questions or 60% of the 100 total questions for each exam section. The only official answer you can get is from the GBCI, but they do NOT give out an exact number. So, different people have different opinions. Here is my justification:

   1) The GBCI intends to use 60% of the correct questions as the benchmark for a passing score. Here is a simple calculation:

      LEED Exams
      Maximum Score: 200
      Minimum Score: 125
      Difference: 200-125 = 75 points (The score difference between someone who answers everything correctly and someone answers everything incorrectly.)

      $75 \times 60\% = 45$ (points)
      $125 + 45 = 170$ points = passing score

      Total questions: 100
      $60\% \times 100 = 60$ = correct answers needed to pass the LEED exams.

   2) The exact number may be different depending on the difficulty of the version of the exam that you are taking.

      Why?

      Because GBCI wants the LEED passing score to be legally defensible. So, it uses subject matter experts to set the minimum level of required LEED knowledge and professional psychometricians to analyze the performance of people taking the beta tests, and uses the **Angoff Method** to decide the final passing score.

      The easier versions of the test will need a higher number of correct answers to pass, and the harder versions of the test will need a lower number of correct answers to pass. So, the end result is the correct number of the questions needed to pass the exam is about 60, but probably NOT exactly 60.

      If you reach the required level of LEED knowledge, no matter which version of the test you take, you should pass.

6. **What are the key areas that the USGBC uses to measure the performance of a building's sustainability?**
Answer: Sustainable Site development (SS), Water Efficiency (WE), Energy and Atmosphere (EA), Materials and Resource (MR), Indoor Environmental Quality (IEQ), Innovations and Design process (ID), and Regional Priority (RP).

Mnemonics: Shall WE Make It Italic and Red?

7. **How many LEED exams does the USGBC have?**
Answer:
1) Before 2009, there were three old versions of the LEED exams: LEED New Construction (LEED-NC) v2.2, LEED Commercial Interior (LEED-CI) v2.0, and LEED Existing Building (LEED-EB) v2.0. You just needed to pass one of the three exams to earn the old title of LEED.
2) In 2009, GBCI started to introduce a new three-tier LEED credential system, and related new versions of LEED exams, including:
    a. The LEED Green Associate Exam. It is also the Part One of ALL LEED AP+ exams.

    b. Part Two of LEED AP+ exams, including the following specialties (exam takers just need to choose one):
    LEED AP BD+C (Building Design and Construction)
    LEED AP ID+C (Interior Design and Construction)
    LEED AP O+M (Operation and Maintenance)
    LEED AP ND (Neighborhood Development)
    LEED AP Homes

8. **Are LEED Exams valid and reliable?**
Answer: Yes. They are valid because they can measure what they intend to measure. They are reliable because they can accurately measure a candidate's skills.

9. **How many member organizations does the USGBC have?**
Answer: More than 11,000 member organizations.

10. **How many regional chapters does the USGBC have?**
Answer: 75.

11. **What is the main purpose of the USGBC?**
Answer: To improve the way a building is designed, built, and used to achieve a healthy, profitable, and environmentally responsible building and environment, and to improve the quality of life.

12. **What are the guiding principles of the USGBC?**
Answer: The USGBC emphasizes not only the decisions themselves but also *how* the decisions are made. Its guiding principles are: advocate the **triple bottom lines** (balance environmental, social and economic needs; some people summarize these as planet, people and profit); build leadership; strive to achieve the balance between humanity and nature; uphold integrity and restore, preserve and protect the environment, species and ecosystem; use democratic and interdisciplinary approaches to ensure inclusiveness to achieve a common goal; openness, honesty and transparency.

13. **How much energy and how many resources do buildings consume in the US?**
Answer: Per the USGBC and US Department of Energy, buildings consume about 39% of total energy, 74% of electricity and 1/8 of the water in the US. Buildings also use valuable land that could otherwise provide ecological resources. In 2006, more than 1 billion metric tons of carbon dioxide was generated by the commercial building sector. That is over a 30% increase from the 1990 level.

14. **What is the most important step to get your building certified?**
Answer: It is to register your building with USGBC online at usgbc.org.

15. **What are the benefits of green buildings and LEED certification?**
Answer: To enhance the building's and company's marketability and to provide branding opportunities, a positive impact on health and the environment. LEED certification may also increase occupants' productivity, reduce building operating costs, and help to create sustainable communities.

16. **Who develops the LEED green building rating systems?**
Answer: The USGBC committee and volunteers.

17. **What current reference guides and specific green building rating systems does the USGBC have?**
Answer: The USGBC has the following reference guides and green rating system portfolio:

1) *The LEED 2009 Reference Guide for Green Building Design and Construction (BD&C)*, which covers the following LEED rating systems:
   - LEED for New Construction (LEED-NC)
   - LEED for Core and Shell (LEED-CS);
   - LEED for Schools
   - LEED for Health Care*
   - LEED for Retails*

2) *The LEED 2009 Reference Guide for Green Interior Design and Construction (ID&C)*, which covers the following LEED rating systems:
   - LEED for Commercial Interiors (LEED-CI)
   - LEED for Retails Interiors*

3) *The LEED 2009 Reference Guide for Green Building Operations and Maintenance (O&M)*, which covers the following LEED rating systems:
   - LEED for Existing Buildings (LEED-EB)
   - LEED for Existing Schools*

Note: These two LEED-EB rating systems are the only systems that cover building operation. All other LEED systems cover building design and construction, but NOT operation.

4) *The LEED for Neighborhood Development Reference Guide (LEED-ND)*.

5) **Application Guide for Multiple Buildings and On-Campus Building Projects (AGMBC)**, 2010 Edition. See link below:
http://www.usgbc.org/campusguidance

6) USGBC has **NOT** updated the following LEED rating systems or related information to LEED 2009 or LEED v3.0:
   - LEED for Homes (**LEED-H**): The new version will be released in 2011.
   - LEED for Laboratories*

*These LEED rating systems are under development or currently in a pilot program. Once their new versions are finalized, the USGBC will sell supplements to their reference guides for them.

The complete sample LEED 2009 credit templates will be available in the future at link below: http://www.usgbc.org/DisplayPage.aspx?CMSPageID=1447

Refer to the following link for various LEED rating systems: http://www.usgbc.org/projecttools

After June 26, 2009, ALL buildings have to be registered under LEED 2009 or v3.0.

18. How does LEED fit into the green building market?
Answer: LEED fits into the green building market by providing voluntary, market-driven, and consensus-based rating systems. It is based on accepted environmental and energy principles, and maintains a balance between emerging concepts and established practices. Green buildings developed under the LEED rating systems can reduce operating costs and create branding and marketing opportunities for buildings and organizations. They are good for the environment and public health, increase occupant productivity and create sustainable communities.

19. What are the benefits of LEED certification for your building?
Answer:
a. The building is qualified for various government initiatives.
b. The building can obtain USGBC (third party) validation of your achievement.
c. You can be recognized for your commitment to environment issues.
d. Branding opportunities and market exposure through media, Greenbuild conference, USGBC, cases studies, etc.

20. What is the procedure of LEED certification for your building?
Answer:
a. Go to www.gbci.org to register your building, the earlier the better. This is the most important step.
b. You must document that your building meets prerequisites and a minimum number of points to be certified.

Refer to the LEED project checklist for the points needed for various levels of LEED certification.

21. How much is the building registration fee and how much is the building LEED certification fee?
Answer: The building registration fee is $450 for USGBC members and $600 for non-members.

The building certification fee varies, but it is $2000 on average.

22. How are LEED credits allocated and weighted?
Answer: LEED credits are allocated and weighted per strategies that will have the greatest positive impacts on the most important environmental factors: CO_2 reductions and energy efficiency. They are weighted against **13 aftereffects of human activities**, including carbon footprint/ climate change (25%), indoor-air quality (15%), resource/fossil-fuel depletion (9%), particulates (8%), water use/water intake (7%), human health: cancer (7%), ecotoxity (6%), eutrophication (5%), land use/habitat alteration (5%), human health: non cancer (4%), smog formation (4%), acidification (3%), and ozone depletion (2%) .

These 13 aftereffects are created by the U.S. Environmental Protection Agency (**EPA**), and are also known as "**TRACI**", a **mnemonic** for "Tool for the Reduction and Assessment of Chemical and Other Environmental Impacts."

1) The USGBC uses a reference building to estimate environmental impact in the 13 categories above.
2) The USGBC also used a tool developed by the National Institute of Standard and Technology (**NIST**) to prioritize the TRACI categories.
3) It also created a matrix to show the existing LEED credits and the TRACI categories, and used data that quantify building impacts on human health and environment to allocate points for each credit.

The points for energy and transportation credits have been greatly increased in LEED 2009, primarily because of the importance of reducing carbon and greenhouse gas emissions. The importance of water efficiency has also been reflected, doubling from 5 to 10 points in some LEED rating systems.

In addition to the weighting exercise, the USGBC also used value judgments because if it only used the TRACI–NIST tool, some existing credits will be worth almost nothing, like the categories for human health and indoor air quality. The USGBC wanted to keep the LEED system somewhat consistent and retained the existing credits including those with few environmental benefits. So each credit was assigned at least one point in the new system.

There are NO negative values or fractions for LEED points.

For instance, a 20% reduction of indoor water-use used to be able to gain points, now this is a prerequisite in LEED 2009.

23. Are there LEED certified products?
Answer: ALL LEED systems certify buildings or projects, NOT products. Products can contribute to a LEED project, and sometimes a product's data is required as part of a LEED submittal package.

Chapter 3
Introduction to the LEED AP ID+C Exam

1. **What is the scope of the LEED Green Associate Exam?**
Answer: Per the USGBC, the content of the LEED Green Associate Exam is limited to:

   a. **Understanding the fundamental credit intents, strategies, requirements, technologies, and submittals for the six major credit categories:** Sustainable site development (SS), water savings/efficiency (WE), energy and atmosphere (EA), materials and resource (MR), indoor environmental quality (IEQ), and innovations and design process (ID), as well as innovation in upgrades, maintenance, and operations.

   b. **Process of LEED application and opportunities for synergy**
   Synergy means the combined effects of two or more agents or forces greater than the sum of their individual.
   The LEED Green Associate Exam may test the following aspects:
   - Site, budget, schedule, program, and other requirements
   - Hard, soft, and life-cycle costs
   - Environmental Building News, **USGBC**, **NRDC** (Natural Resources Defense Council), and other green resources
   - **Green Seal**, **SMACNA** (Sheet Metal and Air Conditioning Contractors National Association Guidelines), **ASHRAE** (American Society of Heating, Refrigeration, and Air-conditioning Engineers), and other standards for LEED credit
   - Waste management, IEQ and energy, and other interaction between LEED Credits
   - **CIR** (Credit Interpretation Requests/Rulings) and previous samples leading to extra points
   - Project registration and **LEED Online**
   - Score Card for LEED
   - Supplementary documentation, project calculations, and other Letter Templates
   - LEED credit **strategies**
   - Property, project, and LEED boundaries
   - Minimum LEED certification program requirements and/or **prerequisites**
   - Certification goal and preliminary rating
   - Opportunities for the same building to get multiple certifications, i.e., commercial interior as well as core and shell
   - Certified building in LEED neighborhood development
   - Operations and maintenance for certified new building construction
   - Requirements for occupancy (e.g., an existing building must be fully occupied for 12 continuous months as described in minimum program requirements)
   - Logo usage, trademark usage, and other USGBC policies

- Requirements for receiving LEED AP credit

c. **Site factors of a project:**
   1) **Connectivity** for Communities
      a) Carts, shuttles, car-sharing membership (e.g. Zipcar™), bike storage, public transportation, fuel efficient vehicle parking, parking capacity, carpool parking, and other means to improve **transportation**
      b) Ramps, crosswalks, trails, and other means to improve **pedestrian access** and circulation

   2) **Zoning requirements** (calculations of site area, floor to area ratio, and other density components, building footprint, open space, development footprint, construction limits, and specific landscaping restrictions)

   3) **Development**
      Heat Islands including albedo, emissivity, non-roof/roof heat island effect; green roofs, Solar Reflectance Index (SRI), etc.

d. **Management of water:**
   1) Understand quality and types of water, i.e., blackwater, graywater, storm water, and potable water

   2) Understand water management, i.e., use low-flush fixtures such as water closets, urinals, sinks, lavatory faucets, and showers to reduce water use; calculations of FTE (Full Time Equivalent); baseline water demand; irrigation; rainwater harvesting

e. **Project system and related energy impact:**
   1) Ozone depletion, chlorofluorocarbon (CFC) reduction, no refrigerant option, fire suppressions without halons or CFC's, phase-out plan, Hydrochlorofluorocarbons (HCFC), and other Environmental Concerns
   2) Green-e providers, off-site generated, renewable energy certificates, and other Green Power

f. **Project materials acquisition, installation and management:**
   1) Commingled, pre-consumer, post-consumer, collection, and other requirements regarding recycled materials
   2) Locally/regionally manufactured and harvested materials
   3) Accounted by weight or volume, written plan, polychlorinated biphenyl (PCB) removal and asbestos-containing materials (ACM) management, reduction strategies, and other requirements regarding construction waste management

g. **Project team coordination, stakeholder involvement in innovation and regional design:**
   1) Civil engineer, landscape architect, architect, heating-ventilation-air-conditioning (HVAC) engineer, contractor, facility manager, and related integrated project team criteria
   2) Building reuse, material lifecycle, and other durability, planning, and management
   3) Appropriate and established requirements, regional green design and construction, and other innovative and regional design measures

h. **Codes and regulations, public outreach and project surroundings:** building, electrical, mechanical, plumbing, fire protection codes, etc.

i. **Ability to <u>support the coordination of team and project</u> and <u>assist</u>** with the process of gathering the necessary requirements and information for the **<u>LEED process</u>** and coordinating the different job functions for **<u>LEED certification</u>**

j. **Ability to <u>support the process</u> of LEED <u>implementation</u>**

k. **Ability to <u>support</u> technical <u>analyses</u> for LEED <u>credits</u>**

2. **What is the scope of the LEED AP ID+C Exam?**
Answer: Part One is exactly the same as the LEED Green Associate Exam. Part Two tests information specific to LEED green building and construction. For more information, refer to the detailed discussion of the exam scope in the LEED AP ID+C exam candidate handbook at:
http://www.gbci.org/main-nav/professional-credentials/resources/candidate-handbooks.aspx

3. **What is the latest version of LEED and when was it published?**

a. **When was LEED v1.0 <u>released</u>?**
Answer: It was first <u>launched</u> in August 1998, but officially <u>released</u> in <u>1999</u>.

b. **When was LEED-NC v2.0 released?**
Answer: It was first <u>published</u> in <u>1999</u>, but officially <u>released</u> in <u>March 2000</u>.

c. **When was LEED-NC v2.1 released?**
Answer: <u>November 2002</u>.

d. **When was LEED-NC v2.2 <u>published</u>?**
Answer: It was first <u>published</u> in <u>2003</u>, but officially <u>released</u> in <u>2005</u>.

e. **When was LEED v3.0 launched?**
Answer: On <u>4/27/2009</u>, the USGBC launched **LEED v3.0**, which includes: a new building certification model, a new LEED Online, and LEED 2009 (Improvement to the LEED rating systems). LEED v3.0 is a part of the continuous evolution of the LEED building rating system. For LEED v3.0, the USGBC is trying to synchronize the prerequisites and credits across each version of the LEED system, and to create a predictable LEED development cycle (similar to other building codes, likely every three years), a transparent environmental/human impact credit weighting (redistributing the available points in LEED), as well as regionalization (regional bonus credits). Refer to the Appendixes of this book for links and more information on LEED v3.0.

4. **How many possible points does LEED v3.0 have?**
Answer: Starting with LEED v3.0 or LEED 2009, **<u>ALL</u>** updated rating systems have or will have **<u>110</u>** possible points, including **<u>100</u>** possible base points and **<u>10</u>** bonus points.

For LEED 2009 for **<u>New Construction and Major Renovations (NC)</u>**, there are <u>26</u> points for Sustainable Sites (SS), <u>10</u> points for Water Efficiency (WE), <u>35</u> points for Energy and Atmosphere (EA), <u>14</u> points for Materials and Resource (MR), <u>15</u> points for Indoor Environmental Quality (IEQ), <u>6</u> bonus points for Innovations in Design (ID), and <u>4</u> bonus points for Regional Priority (RP).

For LEED 2009 for **<u>Core and Shell (CS)</u>**, there are <u>28</u> points for Sustainable Sites (SS), <u>10</u> points for Water Efficiency (WE), <u>37</u> points for Energy and Atmosphere (EA), <u>13</u> points for Materials and Resource (MR), <u>12</u> points for Indoor Environmental Quality (IEQ), <u>6</u> bonus points for Innovations in Design (ID), and <u>4</u> bonus points for Regional Priority (RP).

For LEED 2009 for <u>**Schools**</u>, there are <u>24</u> points for Sustainable Sites (SS), <u>11</u> points for Water Efficiency (WE), <u>33</u> points for Energy and Atmosphere (EA), <u>13</u> points for Materials and Resource (MR), <u>19</u> points for Indoor Environmental Quality (IEQ), <u>6</u> bonus points for Innovations in Design (ID), and <u>4</u> bonus points for Regional Priority (RP).

For LEED 2009 for <u>**Commercial Interiors (CI)**</u>, there are <u>21</u> points for Sustainable Sites (SS), <u>11</u> points for Water Efficiency (WE), <u>37</u> points for Energy and Atmosphere (EA), <u>14</u> points for Materials and Resource (MR), <u>17</u> points for Indoor Environmental Quality (IEQ), <u>6</u> bonus points for Innovations in Design (ID), and <u>4</u> bonus points for Regional Priority (RP).

For LEED 2009 for <u>**Existing Building Operation and Maintenance (EB: O&M)**</u>, there are <u>26</u> points for Sustainable Sites (SS), <u>14</u> points for Water Efficiency (WE), <u>35</u> points for Energy and Atmosphere (EA), <u>10</u> points for Materials and Resource (MR), <u>15</u> points for Indoor Environmental Quality (IEQ), <u>6</u> bonus points for Innovations in Design (ID), and <u>4</u> bonus points for Regional Priority (RP).

LEED for Homes will <u>not</u> be updated until 2011, so its points for various certification levels <u>are still **different** than the LEED 2009 or LEED v3.0 standard</u>:

Certified	<u>45</u>–59	points
Silver	<u>60</u>–74	points
Gold	<u>75</u>–89	points
Platinum	<u>90</u>–136	points
<u>Total</u> possible points	<u>**136**</u>	points

For **LEED for Homes,** there are <u>11</u> points for **Innovations and Design Process (ID)**, <u>10</u> points for **Locations and Linkage (LL)**, <u>22</u> points for Sustainable Sites (SS), <u>15</u> points for Water Efficiency (WE), <u>38</u> points for Energy and Atmosphere (EA), <u>16</u> points for Materials and Resource (MR), <u>21</u> points for Indoor Environmental Quality (IEQ), and <u>3</u> points for **Awareness and Education (AE).**

The pilot version of the **LEED for Retail** rating system was released in July, 2008. It has not been updated to LEED v3.0, so its points for various certification levels <u>are still **different** than the LEED 2009 or LEED v3.0 standard</u>:

Certified	<u>26</u>–32	points
Silver	<u>33</u>–38	points
Gold	<u>75</u>–51	points
Platinum	<u>52</u>–70	points
<u>Total</u> possible points	<u>**70**</u>	points

For the pilot version of **LEED for Retail**, there are <u>16</u> points for Sustainable Sites (SS), <u>5</u> points for Water Efficiency (WE), <u>17</u> points for Energy and Atmosphere (EA), <u>13</u> points for Materials and Resource (MR), <u>14</u> points for Indoor Environmental Quality (IEQ), and <u>5</u> bonus points for Innovations in Design (ID).

The current version of **LEED-NC Application Guide for Multiple Buildings and On-Campus Building Projects (AGMBC)** was released in <u>October 2005</u> and is for use with the LEED-NC Green Building Rating System v2.1 and v2.2. It has <u>not</u> been updated to LEED v3.0 yet, so its points for various certification levels are still **different** than the LEED 2009 or LEED v3.0 standard:

Certified	26–32	points
Silver	33–38	points
Gold	75–51	points
Platinum	52-69	points
Total possible points	**69**	points

For the current version of the **LEED-NC Application Guide for Multiple Buildings and On-Campus Building Projects (AGMBC)**, there are 14 points for Sustainable Sites (SS), 5 points for Water Efficiency (WE), 17 points for Energy and Atmosphere (EA), 13 points for Materials and Resource (MR), 15 points for Indoor Environmental Quality (IEQ), and 5 bonus points for Innovations in Design (ID).

5. **How many different levels of building certification does USGBC have?**
Answer: Four: Certified, Silver, Gold, and Platinum. They are based on the points that a building earns under the LEED green building rating system. For example, based on the checklists provided by the USGBC (See appendixes for links to download these checklists):

No matter which **LEED v3.0** rating system you choose, each LEED v3.0 rating system has **100** base points; a maximum of **6** possible extra points for Innovation in Design, and a maximum of **4** possible extra points for Regional Priority (RP) for each project. There are 6 possible RP points, but you can only pick and choose a maximum of 4 points for each project.

Certified	**4**0–49	points
Silver	**5**0–59	points
Gold	**6**0–79	points
Platinum	**80**	points and above

Refer to the following link to download an Excel file showing the **Regional Priority Credits (RPC)** for all 50 States. With this spreadsheet, you can locate the RPC for your area by zip code:
http://www.usgbc.org/DisplayPage.aspx?CMSPageID=1984

A LEED Project team does not have to do anything special, since LEED 2009 Online will automatically decide which RPC your project will get once you enter the project zip code and other information. If you have more than 4 RPCs, then you need to decide which 4 RPCs you want to use for your project.

The USGBC is working on developing similar RPC incentives for international projects.

6. **What is the process for LEED certification? What are the basic steps for LEED certification?**
Answer: The process and basic steps for attaining LEED certification vary between commercial and residential projects.

LEED certification **process** for **commercial** projects:

1) At the **Beginning Stage**:
   a) Value assessment
   b) Use checklists to decide the preliminary LEED score

c) Select project goals
d) The documentation/assessment of the recommendations for condition treatment
e) **Registration**, i.e. register your project online at www.gbci.org (The most important thing in the first step). The project administrator has access to the CIR database and LEED Online after registration.
f) Identification of project partners
g) Set goals for green building practice
h) Submit incentives applications for tax rebates, federal incentive funds, etc.
i) Start the documentation process

As part of the registration, the LEED project team will have to agree to report post-occupancy water and energy use. This is to allow the USGBC to have a better understanding of the relationship between building performance and LEED credits. There are several ways to achieve the reporting requirements, including signing up for LEED O&M, OR signing a waiver to allow the USGBC to acquire the data applicable directly from the utility company.

2) **Design** Stage:
   a) Assemble documentation for the design phase
   b) Design Submittal
   c) USGBC design phase review
   d) No credit will be awarded at the design submittal. The USGBC will notify you if the credit is denied or anticipated. You can omit this stage and submit everything during the Construction Submittal phase.
   e) You need to finish design submittal before beginning work on the construction documents.

3) **Construction** Stage:
   a) Assemble documentation for the Construction phase
   b) Construction Submittal occurs at completion of the project
   c) USGBC construction phase review
   d) Credit is awarded or denied.
   e) Make sure you submit any changes to the design submittal for USGBC's review and approval.

4) **Certification** of your project:
   a) Award Certification
   b) You will receive an award letter, plaque, and certificate for your building.
   c) You have 30 days to appeal any comments from the USGBC. You need to pay $500 for each appeal.

5) **Misc:**
   a) The fees are lower for USGBC members, and the fees can be paid at different stages (Design Submittal and Construction Submittal, etc). You can also pay the fees online.
   b) You can check your project status online.
   c) USGBC will start to review your project 10 working days after receiving your submittal.

LEED certification process for **homes**:

1) At the **Beginning** Stage:
   a) Contact the LEED for Home provider

Note: A LEED for Homes Provider is the referee when it comes to who is able to be a **Green Rater** for a **LEED** for Homes project. The Provider is responsible for hiring, training, and overseeing the Green Raters. The USGBC requires that each Provider have a quality assurance protocol for its Green Raters. Homebuilders who intend to achieve LEED for Homes certification **must contact a LEED for Homes Provider** organization. Refer to the following link for a complete list of LEED for Homes Provider organizations in the US: http://www.usgbc.org/DisplayPage.aspx?CMSPageID=1554#CA

A Green Rater is a professional who has shown qualifications for energy rating certification, and demonstrates the ability to deal with a home's energy systems and performance. Certification can be obtained from accreditation bodies such as ENERGY STAR, RESNET, BPI, etc.

b) Select project goals
c) **Registration**, i.e. register your project online (The most important thing in the first step). The project administrator has access to the CIR database and LEED Online after registration.
d) Identification of project partners
e) Submit incentives applications

2) **Design** Stage:
   a) Execute preliminary rating
   b) Organize and submit IDs and CIRs
   c) Organize construction documents (CD)
   d) Set up a durability plan
   e) Set up contracts including the scope of work for various trades

3) **Construction** Stage:
   a) **Green rater** carries out thermal bypass inspection
   b) Green rater carries out optional durability inspection
   c) Various trades sign off on finished tasks
   d) Fill out accountability form

4) **Verification** Stage:
   a) Green rater finishes verification
   b) Provider review of green rater's work
   c) Provider submits documentation to the USGBC

5) **Certification** of you project:
   a) USGBC review
   b) "Denied" or "Achieved"
   c) You have 30 days to appeal any comments from the USGBC. You need to pay $500 for each appeal.

   If **achieved**:
   a) Award Certification
   b) You will receive an award letter, plaque, and certificate for your building.

6) **Misc:**
   a) The fees are lower for USGBC members, and the fees can be paid at different stages (Design Submittal and Construction Submittal, etc). You can also pay the fees online.
   b) You can check your project status online.
   c) USGBC will start to review your project 10 working days after receiving your submittal.

7. What does the registration form include?
Answer: The registration form includes account login information, project contact information, project details (type, title, address, owner, tenant, gross square footage, budget, site condition, current project phase, scope, occupancy, etc.), and the confidentiality status of the project.

8. What is precertification?
Answer: Precertification is for the Core and Shell program only. It is a formal recognition of a project where the owner intends to seek Core and Shell certification, and gives the owner/developer a marketing tool to attract financers and tenants. A project needs to meet a scorecard requirement for precertification. Precertification is granted after an early design review by the USGBC, and the review usually takes less than one month. Precertification costs $2,500 for USGBC members and $3,500 for non-members. It does NOT cover the fee for certification, and it does NOT guarantee the final certification of the project.

9. What is a CIR?
Answer: CIR stands for Credit Interpretation Request or Credit Interpretation Rulings, depending upon its context.

10. When do you submit a CIR?
Answer: When there are conflicts between two credit categories or the USGBC Reference Guide does not give you enough information. The building registration fee used to cover two free CIRs, but after 11/15/05, it costs $200 to submit each CIR.

11. What are the steps for submitting a CIR?
Answer:
1) Check your projects against each LEED credit or prerequisite.
2) Check the USGBC Reference Guide.
3) Review the GBCI and USGBC websites for previous CIRs.
4) If you cannot find a similar CIR, then submit a new CIR to GBCI online.
5) Do not mention the contact information, name of the credit, or confidential information.
6) Submit only the essential and required information, and do not submit it in a letter format.
7) Submit one CIR for each prerequisite or credit.
8) Do not include any attachments.
9) Include details and background information, 600 words max (4,000 characters including spaces).

12. Will submitting a CIR guarantee a credit?
Answer: No, a CIR only provides feedback, and it will not guarantee a credit. Also, no credit will be awarded in the CIR process.

13. Which tasks are handled by the GBCI and what tasks are handled by the USGBC?
Answer: In 2008, the Green Building Certification Institute (GBCI) spun off from the USGBC.

The USGBC still handles the online tool, LEED rating system development, and related educational offerings, etc.

The GBCI took over some responsibilities from the USGBC and handles building LEED certification and LEED professional credentialing.

For building LEED certification, the GBCI oversees 10 organizations including Lloyd's Register Quality Assurance and Underwriters Laboratories, which manage the project review process.

This separation of the tasks for the USGBC and the GBCI will meet the protocols of the American National Standard Institute (ANSI) and International Organization for Standardization (ISO), and will make the building certification a true third-party process.

14. What are MPRs?

Answer: MPRs are LEED's **Minimum Program Requirements**. A project must meet MPRs to qualify for LEED certification.

MPRs serve three goals:

1) Clear guidance for customers
2) Maintain LEED program integrity
3) Make the LEED certification process easier

MPRs include some very basic requirements, like:

1) The building must be 1,000 s.f. minimum for LEED-NC, LEED Schools, LEED-CS, and LEED-EB: O&M, and 250 s.f. minimum for LEED-CI.
2) A LEED project must be a permanent and complete space or building.
3) A LEED project must comply with environmental laws
4) A LEED project must have a reasonable site boundary
5) The building to site ratio must be 2% or higher.
6) A LEED project must commit to sharing water and whole-building energy usage data
7) A LEED project must meet a minimum occupancy rate

Refer to the link below for detailed information:
http://www.usgbc.org/projecttools

15. Which types of projects should use LEED-NC?

Answer: LEED-NC should be used for new construction or major renovation of commercial occupancies, like a residential building of 4 or more habitable floors, hotels, offices, or institutional buildings (churches, museums, libraries).

A **major renovation** means major building envelope changes, significant HVAC renovation and interior rehabilitation.

If a building project does not involve major renovation, then it is better to seek LEED-EB: O&M certification.

16. Which types of projects should use LEED-CS?

Answer: Use LEED-CS for projects where the developer controls the core and shell (building envelope, HVAC and fire protection systems), but has no control over tenant improvements. These projects may include retail centers, lab facilities, warehouses, medical office buildings, or commercial office buildings. If the developer occupies **50% or less** of the building's **leasable area,** then he can also seek LEED-CS certification.

17. Which types of projects should use LEED for Schools?
Answer: Use LEED for Schools on new construction or major renovation of school buildings.

Academic buildings in K-12 schools must use LEED for Schools.

Prekindergarten and post-secondary academic buildings can use either LEED-NC or LEED for Schools.

Non-academic campus buildings like dormitories, maintenance facilities or administrative buildings can use either LEED-NC or LEED for Schools.

If a campus project does not involve new construction or major renovation, then LEED-EB: O&M is often most applicable.

18. Which types of projects should use LEED-CI?
Answer: Use LEED-CI for Commercial Interior projects, i.e., commercial tenant improvement projects.

Chapter 4

LEED AP ID+C Exam Technical Review

In this chapter, we introduce the LEED certification process, key components of the LEED rating system, as wells as the purpose, core concepts, strategies, incentives, recognitions, and regulations for each LEED credit category.

1. **What do green buildings address?**
   Green buildings mitigate the effects of the degradation of ecosystems/habitats and resource depletion, reduce the costs of operating and owning living and working spaces, improve occupant productivity and comfort, improve indoor environmental quality, and reduce water consumption.
   Mnemonics: DR. COIN C. See underlined letters in the sentences above.

2. **Key stake holders and an integrative approach**
   Green buildings employ an **integrative approach** and encourage the participation of **key stake holders**, including the:

   1) **Client:** Facilities Management Staff, Facilities O&M Staff, Community Members, Owner, and Planning Staff. **Mnemonics**: MOM COP

   2) **Design team:** Civil Engineer, Landscape Architect, Architect, Mechanical and Plumbing Engineer, Electrical Engineer, Structural Engineer, Commissioning Authority, and Energy and Daylighting Modeler.
   Mnemonics: CLAMPES CAM

   3) **Builder:** General Contractors, EMP Subcontractors, Cost Estimator, Construction Manager, and Product Manufacturer.
   Mnemonics: G EMP EMP

   There are many **benefits** of an **integrative approach**, including better indoor air quality, improved occupant performance, reduced operating and maintenance costs, durable facilities, reduced environmental impact, potentially no increase in construction cost, optimized return on investment, and opportunity of learning.
   Mnemonics: bird in pool

3. **The Mission of the USGBC**
   The **mission** of the USGBC is:
   1) To improve quality of life
   2) To change the way we design, build, and operate communities and buildings
   3) To create a healthy, prosperous, socially, and environmentally responsible environment

4. The structure of LEED rating system

In addition to the main credit categories of SS, WE, EA, MR, IEQ, ID, and RP, the following factors are very important to the LEED rating systems: linkages and locations, awareness and education, green buildings and infrastructure, as well as neighborhood design and pattern.

The structure of the LEED rating system includes three tiers:

1) The main credit categories of SS, WE, EA, MR, IEQ, ID, and RP
2) Each main credit category includes prerequisites and credits
3) Each prerequisite or credit includes intents and requirements (or paths to achieve LEED points)

5. LEED certification tools

LEED certification tools include:

1) USGBC Reference Guides
2) LEED Rating systems
3) LEED Online
4) LEED Scorecard
5) LEED Letter Template
6) CIRs (Credit Interpretation Rulings)
7) LEED Case Studies
8) USGBC and GBCI websites:
   www.USGBC.org
   www.GBCI.org

The GBCI requires that the project team submit an overall narrative and complete the LEED Online documentation for the LEED certification application. The general documentation includes the project timeline and scope, site conditions, team identification, usage data, and occupant information, etc.

The project's overall narrative includes the team, building, site, and applicant's organization.

LEED Online includes the project's detailed information and template/complete documentation requirements for completing prerequisites and credits: my Action Items, potential LEED ratings, attempted credit summary (not awarded, anticipated, denied, and total attempted), appealed credit summary, and credit scorecard, etc. It also includes embed tables and calculators to assure accuracy and completeness of the submittal package.

LEED Online also includes the definition for **declarant**, Licensed Professional Exemption Form and related information. A **declarant** is the team member(s) who signs off on the documents and indicates who is responsible for each credit or prerequisite.

A licensed professional exemption form is the form for a team member who is a licensed/registered landscape architect, architect, or engineer to use as a tool to request waiver for eligible submittal requirements. Licensed professional exemptions are shown in the related LEED credit section of LEED Online.

With LEED Online, project teams can upload support files, submit applications for reviews and CIRs, receive feedback from the reviewer, contact customer service, generate specific reports for the project, and obtain project LEED certification, as well as gain access to additional LEED resources such as tutorials, FAQs, sample documentation, and offline calculators.

If you have multiple projects, you can access all of them via LEED Online.

The GBCI also issues LEED certificates for successful projects via LEED Online.

The credit scorecard includes the project name, address, and a list of all points in various credit categories. With LEED Online, you can chose to collapse all credit categories or view a printer friendly scorecard.

The LEED letter template includes your name, your company name, specific and credit-related project information such as the project site area (s.f.), gross building area (s.f.), and the credit path you choose for each credit.

As a LEED professional, you also need to know the **Synergies** between the main credit categories of SS, WE, EA, MR, IEQ, ID, and RP. Almost every LEED exam will have a large percentage of test questions related to **Synergies.**

6. Specific Technical Information

A. Sustainable sites (SS)

Overall purpose:

1) **Construction activity pollution prevention:** prevent soil erosion by wind or storm water runoff during construction; protect and stockpile topsoil for reuse; avoid waterway sedimentation; prevent airborne dust and particle matter from polluting the air.
2) **Site selection: develop proper sites:** locate a building wisely on a site to alleviate environmental impact.
3) **Development density and community connectivity: reuse existing sites and/or buildings:** encourage development in urban areas with existing infrastructure.
4) **Brownfield redevelopment: restore and/or protect sites:** rehabilitate damaged sites with environmental contamination; restore damaged habitats to promote biodiversity.
5) **Alternative transportation: reduce automobile use:** reduce land development impact and pollution.
6) **Site development: protect or restore habitat and maximize open space; protect prime farmland and undeveloped land/natural areas** (habitat and greenfields) and natural resources; promote biodiversity via a high ratio of open space to development footprint.
7) **Storm water design: quality control and quantity control:** manage storm water runoff; increase on-site infiltration; reduce impervious cover; eliminate contaminants; eliminate storm water run-off pollution; limit disruption of natural water hydrology/natural water flows; capture and process storm water runoff with the help of a storm water management plan and acceptable **BMPs (best management practices)**.

8) **Heat island effect: non-roof and roof:** reduce heat islands' impact on wildlife habitat, humans and microclimate. Heat island refers to the extra thermal gradient in developed areas when compared with undeveloped areas.
9) **Light pollution reduction:** reduce development impact on nocturnal environments; reduce glare and improve nighttime visibility, and reduce sky-glow and light trespass from site and building to increase night sky access.

Mnemonics: Carole Smith Does Business As "So Sweet, Hot Love." (See underlined letters above also).

Core concepts:

1) **Location/Transportation**
   a) Reduce demand for transportation
   b) Improve the efficiency of transportation
2) **Selection of site**
   a) Develop in existing high density areas
   b) Brownfield redevelopment
   c) Connectivity and site selection
3) **Management and design of the site**
   a) Stewardship of the site
   b) Site development
   c) Light pollution reduction
   d) Pest management integration
4) **Management of storm water**
   a) Storm water quantity reduction and water quality protection
   b) Impervious surface impact

Recognition, regulation and incentives:

1) **Financial incentives**

Local, State, and Federal government programs

- Encourage the reuse of "brownfield" sites or infill
- Promote water quality protection with the help of combining low impact development and smart growth

Overall strategies and technologies:

Note: Not **all** strategies or technologies have to be used simultaneously for your project.

1) Development density (use existing infrastructure) OR
2) Community connectivity:
   - Consider diverse land uses and provide pedestrian access
   - Provide access to recreational and public spaces
   - Plan walkable, safe connections between buildings, surrounding developments, and open spaces
3) Develop in previously developed locations
4) Use telecommuting and flexible work schedules to reduce transportation demand
5) Support efficient modes of transportation
6) Provide access to transit
7) Provide bicycle accessibility
8) Use efficient vehicles

9) Support car/vanpools
10) Protect or restore habitat during the site development process
11) Restore and/or protect open spaces:
    - Plan for protected areas on-site
    - Plan for easements and protected areas off site
12) Manage and intercept storm water
13) Apply cool roof technologies
14) Reduce the duration of lighting use and lighting density, also use light fixtures that comply with dark sky requirements

Specific Technical Information:

SSc1: Site Selection (1-5 points)

Purpose:

To encourage tenants to choose buildings that avoid inappropriate sites and/or are wisely located on a site to alleviate environmental impact.

Credit Path:

Ch1 (Choice One): Choose a building that is LEED Certified (5 points).

Related Strategies and Technologies for Base Building:

1) Create an **Erosion and Sedimentation Control (ESC) Plan** early in the project, preferably during the design stage.
2) Use the **Stabilization Method**: mulching, temporary, or permanent seeding.
3) Use **Structural Method** if erosion has already occurred: silt fencing, earth dikes, sediment basins, and sediment traps.

Your base building should avoid sites that meet any one of the following criteria:

a) Virgin (previously undeveloped) land that is lower than five feet above the elevation of the 100-year flood per the definition of Federal Emergency Management Agency (FEMA).
b) Virgin land that is within 50 feet of a water body, such as a lake, river, stream, or sea (NOT including small man-made ponds), per the terminology of the Clean Water Act.
c) Within 100 feet of wetlands per the definition of the US Code of Federal Regulations, 40 CFR, Parts 230-233, and Part 22, Definition of Wetlands, and areas of special concern or isolated wetlands per local rules, OR within setback distance from wetlands per local codes, whichever is more restrictive.
d) Habitat for any species on state OR federal endangered or threatened lists.
e) Public parkland unless a public land owner has accepted a trade for land of equal or greater value as parkland (This does not apply to Park Authority projects).
f) Prime farmland per the definition of the US Department of Agriculture (USDA) in the US Code of Federal Regulations (CFR).

Ch2: Place the tenant space in a building that has one or more of the following features at submittal (1-5 points). Each of the following features can be achieved by meeting the related LEED 2009 NC credit:

SSc1, Ch2F1 (Choice Two, Feature 1): Brownfield Redevelopment (1 point)

Note: The USGBC reference uses the word **path**, but **feature** is a clearer choice given the context of the credit.

1) A building located on a contaminated site as documented by a local voluntary cleanup program or by an ASTM E1903-97 Phase II Environmental Site Assessment;

OR

2) A building located on a brownfield as defined by a federal, state, or local governing agency.

The site contamination must have been effectively remediated.

SSc1, Ch2F2: Storm Water Design: Quantity Control (1 point)

If the existing imperviousness of a building (prior to development) is equal to or less than 50%:

1) The building must have implemented a storm water management plan to prevent post-development peak quantity and discharge rate from exceeding that of pre-development for the 1-1/2 year, 24-hour design storms.

OR

2) The building must have applied a quantity control strategy, a stream channel protection strategy, and a storm water management plan to prevent excessive erosion from reaching receiving stream channels.

If the existing imperviousness of a building (prior to development) is greater than 50%:

3) The building must have implemented a storm water management plan to decrease the storm water runoff volume from the 1-1/2 year, 24-hour design storms by 25%.

The LEED project team can achieve the mitigation in several ways: harvesting and reusing rainwater, including retention ponds for storm water, improving site imperviousness, etc.

The LEED project team should use the actual local rainfall for storm water values, OR use the 10-year annual average if the actual rainfall exceeds the 10-year annual average.

SSc1, Ch2F3: Storm Water Design: Quality Control (1 point)

1) A building that promotes infiltration, reduces impervious paving, as well as captures and processes storm water runoff from the average annual rainfall of all storms not exceeding the level of the 2-year, 24-hour storm, with the help of storm water treatment systems, and acceptable **BMPs (best management practices).**

AND

2) BMPs shall be able to remove 80% of the average annual post-development **total suspended**

solids (TSS) load per an existing monitoring report and 40% of the average annual **total phosphorus (TP)** of the site. The BMPs must meet the most restrictive one of the two following criteria:

a) Comply with specifications and standards from a local or state program;

OR

b) Comply with Chapter 4, Part 2: Urban Runoff, of the EPA Guidance Specifying Management Measures for Sources of Nonpoint Pollution in Coastal Waters, January 1993 (EPA 840B92002).

SSc1, Ch2F4: Heat Island Effect: Non-Roof (1 point)

A building that meets the following criteria:

1) Uses an open grid pavement system (50% pervious minimum) for 50% of parking lots.

OR

2) Uses paving materials with a SRI (Solar Reflectance Index) of 29 or more; and/or provides shade from tree canopies (within five years of occupancy). Trees shall be planted at the time of occupancy; and/or provides open grid pavement system that individually or in total equal to 30% of the impervious surfaces of the site.

OR

3) Provide cover/roof for 50% or more of the parking spaces. The qualified parking space undercover can be under buildings, decks, roofs, or underground.

SSc1, Ch2F5: Heat Island Effect: Roof (1 point)

A building that meets the following criteria:

1) For at least 75% of the roof surface, use roofing materials with a SRI of 78 or more for low-sloped roof (Slope 2:12 or less) and a SRI of 29 or more for steep-sloped roof (Slope more than 2:12).

Roofing materials with a SRI lower than the ones listed above may be used if they meet the following:

$$\frac{SRI\ of\ Installed\ Roof}{Required\ SRI} \times \frac{Area\ of\ Roof\ Meeting\ Minimum\ SRI}{Total\ Roof\ Area} \geq 75\%$$

OR

2) Use a vegetated roof for 50% or more of the roof area.

OR

3) Use a combined vegetated roof and high albedo (reflectivity) surfaces that meet the following:

$$\frac{Area\ of\ Vegetated\ Roof}{0.5} + \frac{Area\ of\ SRI\ Roof}{0.75} \geq Total\ Roof\ Area$$

AND

Use roof materials with a SRI of 78 or more for low-sloped roof (Slope 2:12 or less) and a SRI of 29 or more for steep-sloped roof (Slope more than 2:12).

SSc1, Ch2F6: Light Pollution Reduction (1 point)

A building that meets the following criteria:

1) For Interior Lighting:
   Use automatic controls (automatic sweep timers, programmed master lighting control panels, or occupancy sensors) to turn off 50% of non-emergency interior lights with a direct line of sight to any openings in the building envelope between 11 p.m. to 5 a.m. Provide occupant sensing devices or capacity for manual override during after-hours use. The override shall not last more than 30 minutes.

OR

2) All any non-emergency interior lights that have a direct line of sight to openings in the building envelope should have shielding (closed/controlled by automatic device between 11:00 p.m. and 5:00 a.m.) to allow for less than 10% transmittance.

   Buildings that operate 24 hours a day shall use option 2).

SSc1, Ch2F7: Water Efficient Landscaping: Reduced by 50% (Outdoor Water, 2 points)

A building that meets the following:

1) Based on calculated mid-summer (July) baseline case (conventional use), reduce 50% of the potable consumed by irrigation.

The building can achieve this water reduction from any combination of the following:

a) Capture and use rainwater.
b) Recycle wastewater.
c) For non-potable irrigation use, use water treated and delivered by a public agency.
d) High efficient irrigation.
e) Choose the right plant species.
f) Microclimate and density factor.
g) The building owner can pump groundwater seepage away from the vicinity of building foundations and slabs and use it for irrigation, but you must demonstrate doing so does not affect site storm water management systems.

SSc1, Ch2F8: Water Efficient Landscaping: No Potable Water Use or No Irrigation (Outdoor Water, 2 points in addition to Feature 7)

A building that meets the following criteria:

1) Use plants that do not need a permanent irrigation system. The building owner can use temporary irrigation systems for plant establishment only if he removes them within one year.

OR

2) Do not use any potable water. For non-potable irrigation use, the building owner can only use recycled graywater, recycled wastewater, captured rainwater, or water treated and delivered by a public agency.

SSc1, Ch2F9: Innovative Wastewater Technologies (Indoor Water, 2 points)
A building that meets the following criteria:
1) Use non-potable water (recycled graywater, captured rainwater, and municipally treated wastewater) or water-saving plumbing fixtures (urinals, toilets) to reduce 50% of the municipally provided potable water use for building sewage conveyance.

OR
2) Follow tertiary standards and treat 100% of the on-site wastewater. The building owner can use a living machine/natural system or a packaged mechanical wastewater treatment system.

SSc1, Ch2F10: Water Use Reduction: 30% Reduction (1 point)
A building that meets the following criteria:
1) Reduce water use for the whole building by 30% and has an ongoing plan to require future tenants to comply with.

SSc1, Ch2F11: On-Site Renewable Energy (1 to 2 points)
A building that meets the following criteria:
1) Provides a minimum of 2.5% (1 point) or 5% (2 point) of the entire building's energy use (percentage of annual energy cost) with the help of on-site renewable energy systems.

SSc1, Ch2F12: Other Quantifiable Environmental Performance (1 point)
A building that provides other quantifiable environmental benefits at time of choosing the tenant space will qualify for this credit.

Detailed discussion of the Choices and features of SSc1: Site Selection

SSc1, Ch1: Choose a building that is LEED Certified (5 points)

Considerations and Benefits:

1) **Environmental:** In the U.S.:
Per the U.S. DOE (Dept. of Energy), buildings consumed 72.4% of the total electricity produced in 2006; per the EIA (Energy Information Administration), buildings emitted 38% of the total CO_2 released in 2008; per the USGS (the US Geotechnical Survey), buildings use 13.6% of the total potable water available, which totaled 15 trillion gallons in 2000.

   A building certified under the LEED CS, NC, Schools or O+M rating system indicates the building owners have taken important measures to conserve resources, protect biodiversity and ecosystems, and improve IEQ (indoor environmental quality) for building occupants.

2) **Financial:** LEED-certified buildings have been proven to provide economic benefits to tenants, including improved employee productivity as well as reduced operation and maintenance cost.

Tenants can weigh the increased lease value against the reduced costs before making their final decision.

Submittals:

1) Documentation of the base building's LEED certification from the building manager or owner.

Synergies:

1) Selecting a building that is LEED certified will increase the likelihood of obtaining additional credits in various categories for the LEED CI rating system.

Possible Strategies and Technologies:

1) Make it a priority to select a LEED certified base building.
2) Obtain the documentation of the base building's LEED certification at an early stage. It will help you identify base building systems and related LEED credits, and simplify the LEED certification process for LEED CI.
3) As early as possible, inform real estate agents that you are working with that locating tenant space in a LEED certified base building is an important criterion.
4) Use the USGBC website and local chapters to locate LEED certified base buildings or buildings seeking LEED certification.

Extra Credit (Exemplary Performance):

None

Project Phase:

Pre-Design

LEED Submittal Phase:

Design

Related Code or Standard:

None

Responsible Parties:

The tenant and the real estate agent

SSc1, Ch2F1: Brownfield Redevelopment (1 point)

Considerations and Benefits:

1) **Environmental:**
   Per the EPA, the U.S. has over 450,000 brownfields. It is possible to stabilize and remove hazardous wastes from the project sites' groundwater and soil, and reduce their impacts on wildlife and humans. This can help to revitalize disadvantaged or depressed areas.

2) **Financial:** Brownfield sites often have opportunities for tax incentives, government grants, and the advantages of availability of labor and existing infrastructure. Buildings in brownfield sites can have lower lease rates.

Submittals:

1) Documentation of the previous site contamination and related remediation efforts.

Synergies:

None

Possible Strategies and Technologies:

1) Choose a base building located in a remediated brownfield.
2) Local, state, or federal agencies may have information or documentation relating to the brownfield sites and related remediation efforts.
3) As early as possible, inform real estate agents that you are working with that locating tenant space in a LEED certified base building is an important criterion.

Extra Credit (Exemplary Performance):

None

Project Phase:

Pre-Design

LEED Submittal Phase:

Design

Related Code or Standard:

1) ASTM E1903-97 Phase II Environmental Site Assessment, effective 2002
2) U.S. EPA, Definition of Brownfields
3) Comprehensive Environmental Response, Compensation and Liability Act (CERLA): Super Fund

Responsible Parties:

The tenant and the real estate agent

SSc1, Ch2F2: Storm Water Design: Quantity Control (1 point)

Considerations and Benefits:

1) **Environmental:**

Storm water causes significant pollution in all types of water bodies. Impervious surfaces (parking lots and roads) and soil compaction increase storm water runoff containing contaminants and sediment like atmospheric deposition, fertilizers, mechanical equipment waste, pesticides, and vehicle fluid leaks. The increased storm water can overload sewers and pipes, affect recreation and navigation, and lower water quality. It also increases the cost and maintenance of infrastructure needed to treat and convey runoff.

Storm water volumes and velocities have significant impact on the health of streams. Their increases can cause severe erosion and bankfull events, widen channels, and stream **downcutting** (vertical erosion). LEED on-site management practices allow storm water to infiltrate the ground, recharge natural aquifer, restore base flows for streams, and reduce the intensity and volume of storm water flows. Choosing a base building that meets SSc1, Ch2F2 indicates the tenant's commitment to reduce storm water and the related environmental benefits.

Submittals:

1) Provide storm quantities and rates for pre- and post-development conditions.
2) Provide a storm water plan assessment prepared by a civil engineer or other design professional.
3) Provide storm water management strategies and the percentage of rainfall that each strategy is handling.

Synergies:

Reducing the quantity and rate of storm water runoff may contribute to the following credit:

- SSc1, Ch2F3: Storm Water Design: Quality Control

Decreasing impervious surfaces and harvesting rainwater may contribute to the following credits:

- SSc1, Ch2F4: Heat Island Effect: Non-Roof
- SSc1, Ch2F7: Water Efficient Landscaping: Reduced by 50%
- SSc1, Ch2F8: Water Efficient Landscaping: No Potable Water Use or No Irrigation
- SSc1, Ch2F10: Water Use Reduction: 30% Reduction

Possible Strategies and Technologies:

Choose a building that meets SSc1, Ch2F2, OR some or all of the following criteria:

1) Promote infiltration and maintain natural storm water flows.
2) Design pervious paving, vegetated roof, etc., to reduce impervious surface.
3) Reuse storm water for non-potable water uses such as urinal and toilet flushing, landscape irrigation and custodial needs.
4) Guidelines for capturing and reusing storm water runoff: when and how it will be used? Drawdown, drainage area, conveyance system, pretreatment, and pressurization.
5) Treatment ponds and underground facilities.

The local governing agencies may have a storm water management permit application and related information on the LEED storm water management for the base building.

Extra Credit (Exemplary Performance):

None

Project Phase:

Pre-Design

LEED Submittal Phase:

Design

Related Code or Standard:

None

Responsible Parties:

The tenant and the real estate agent

SSc1, Ch2F3: Storm Water Design: Quality Control (1 point)

Considerations and Benefits:

1) **Environmental:**

   Storm water causes significant pollution in all types of water bodies. Impervious surfaces (parking lots and roads) and soil compaction increase storm water runoff containing contaminants and sediment such as atmospheric deposition, fertilizers, mechanical equipment waste, pesticides, and vehicle fluid leaks. The increased storm water can overload sewers and pipes, affect recreation and navigation, and lower water quality. It also increases the cost and maintenance for infrastructure needed to treat and convey runoff.

   Storm water volumes and velocities have significant impact on the health of streams. Their increases can cause severe erosion and bankfull events, widen channels, stream **downcutting** (vertical erosion), reduce **biological diversity**, and damage **aquatic habits**. LEED on-site management practices allow storm water to infiltrate the ground, recharge natural aquifer, restore base flows for streams, and reduce the intensity and volume of storm water flows. Choosing a base building that meets SSc1, Ch2F3 indicates the tenant's commitment to reduce storm water and the related environmental benefits.

Submittals:

1) A list of **BMPs** to treat storm water, a description of filtration contribution of each measure, as well as the percentage of rainfall treated by each measure.
2) For structural controls, a list and a description of the pollutant removal performance of each measure as well as the percentage of rainfall treated by each measure.

Synergies:

Reducing the quantity and rate of storm water runoff may contribute to the following credit:

- SSc1, Ch2F2: Storm Water Design: Quantity Control

Decreasing impervious surfaces and harvesting rainwater may contribute to the following credits:

- SSc1, Ch2F4: Heat Island Effect: Non-Roof
- SSc1, Ch2F7: Water Efficient Landscaping: Reduced by 50%
- SSc1, Ch2F8: Water Efficient Landscaping: No Potable Water Use or No Irrigation
- SSc1, Ch2F10: Water Use Reduction: 30% Reduction

Possible Strategies and Technologies:

Choose a building that meets SSc1, Ch2F3, or some or all of the following criteria:

1) Promote infiltration and reduce imperviousness to reduced pollutant loading.
2) Use **nonstructural** techniques such as vegetated swales, raingardens, rainwater recycling, and disconnecting imperviousness.
3) Use **structural** measures: manhole treatment devices, rainwater cisterns, and ponds.
4) Use **alternative** surfaces like pervious paving, vegetated roof, or grid paver, etc.
5) Use vegetated filters, open channels, and wetlands or other integrated **natural and mechanical treatment systems** to treat storm water runoff.
6) Apply Environmentally Sensitive Design, Low Impact Development, or other **sustainable design strategies**.

Note: there are three distinct climates in the United States. Depending on how much annual rainfall an area receives, it can be classified as an:

- **Arid watershed:** 20 inches or less rainfall annually
- **Semi-arid watershed:** 20 to 40 inches of rainfall annually
- **Humid watershed:** 40 inches or more of rainfall annually.

For this feature, treating runoff from 90% of the average annual rainfall equates to:

a) Arid watersheds: 0.5 inches of rainfall.
b) Semi-arid watersheds: 0.75 inches of rainfall.
c) Humid watersheds: 1 inch of rainfall.

Local governing agencies may have a storm water management permit application and related information on the LEED storm water management for the base building. Tenants can consult the design documents, facility personnel, code officials, or manufacturers for information on the base building's storm water treatment system. Tenants can also refer to manufacturer's cutsheets to confirm the installed system can remove phosphorus and suspended solids per referenced standards.

Extra Credit (Exemplary Performance):
None

Project Phase:
Pre-Design

LEED Submittal Phase:
Design

Related Code or Standard:

1) BMP: Best Management Practice
2) EPA: Environmental Protection Agency:
   a) U.S. EPA 840B92002, Guidance Specifying Management Measures for Sources of Non-Point Pollution in Coastal Waters, effective January.
   http://www.epa.gov/owow/nps/MMGI
   b) Hardcopy or microfiche (836 pages): National Technical Information Service (PB93-234672)
   http://www.ntis.gov

Responsible Parties:
The tenant and the real estate agent

SSc1, Ch2F4: Heat Island Effect: Non-Roof (1 point)

Considerations and Benefits:

1) **Environmental:**
   Non-reflective, dark surfaces of roofs, walkways, and parking absorb heat from the sun, radiate it into adjacent areas, and contribute to the heat island effect. Heat island effect can raise the temperature in urban areas by anywhere from 2° to 4° F compared to similar rural or suburban areas. This phenomenon increases the demands on HVAC systems and for electricity supply, and subsequently the related pollution and greenhouse gases. It damages wildlife migration corridors

and habits. It creates a larger difference between nighttime and daytime temperatures, which can have negative impacts on animals and plants.

2) **Financial:** Over the long term, HVAC operating costs can be significant. Choosing a base building that meets SSc1, Ch2F4 can save the tenant a substantial amount of money in HVAC operating costs.

Submittals:

1) **For shaded condition:** A site plan highlighting all nonroof hardscape. Label hardscape used for credit and list the SRI for the compliant surface.
2) **For covered parking condition:** Show the number of spaces and portion covered, and the SRI for roofs covering the parking areas.
3) **For open-grid paving condition:** A site plan highlighting all open-grid paving. Documentation for open-grid system used.

Synergies:

Open-grid paving system may contribute to the following credits:

- SSc1, Ch2F2: Storm Water Design: Quantity Control
- SSc1, Ch2F3: Storm Water Design: Quality Control

If base buildings have shade provided by vegetation, refer to the following credits for landscape irrigation criteria:

- SSc1, Ch2F7: Water Efficient Landscaping: Reduced by 50%
- SSc1, Ch2F8: Water Efficient Landscaping: No Potable Water Use or No Irrigation

Possible Strategies and Technologies:

Choose a building that meets SSc1, Ch2F4, or some or all of the following criteria:

1) Use high-reflectance materials for hardscape and provide shaded surfaces with landscaped features.
2) Use open grid paving, vegetated roofs or high-albedo (high-reflectivity) materials to replace conventional surfaces like sidewalks, roads and roofs, etc., to reduce heat absorption.

SRI (Solar Reflectance Index) refers to a built surface's capacity to reflect solar heat in terms of a small rise in temperature. SRI for a standard white surface (emittance 0.90, reflectance 0.80) is 100, SRI for a standard black surface (emittance 0.90, reflectance 0.05) is 0. A material's emittance value and reflectance value can be used to calculate its SRI. **SRI** is calculated per ASTM E 1980-01. Emittance is calculated per ASTM C 1371 or ASTM E 408. **Reflectance** is calculated per ASTM C1549, ASTM E903 or ASTM E1918. Standard SRI values can be found in the related USGBC reference guide. Product information can also be obtained at www.coolroofs.org (Cool Roof Rating Council website).

Extra Credit (Exemplary Performance):

The project may earn extra points under SSc1, Ch2F12: Other Quantifiable Environmental Performance if the tenant can show the building has met 2 or more of the credit paths in SSc1, Ch2F4.

Project Phase:

Pre-Design

LEED Submittal Phase:

Design

Related Code or Standard:

None

Responsible Parties:

The tenant and the real estate agent

SSc1, Ch2F5: Heat Island Effect: Roof (1 point)

Considerations and Benefits:

1) **Environmental:**
   Non-reflective, dark surfaces of roofs, walkways, and parking absorb heat from the sun, radiate it into adjacent areas, and contribute to the heat island effect. Heat island effect can raise the temperature in urban areas by anywhere from 2° to 4° F compared to similar rural or suburban areas. This phenomenon increases the demands on HVAC systems and for electricity supply, and subsequently the related pollution and greenhouse gases. It damages wildlife migration corridors and habits. It creates a larger difference between nighttime and daytime temperatures, which can have negative impacts on animals and plants.

2) **Financial:** Over the long term, HVAC operating costs can be significant. Choosing a base building that meets SSc1, Ch2F4 can save the tenant a substantial amount of money in HVAC operating costs.

Submittals:

1) A roof plan showing the total roof area and the vegetated roof area and/or the highly reflective roof area.
2) A list of roofing products used on the project and their SRI, slopes, reflectance percentage, emittance percentage, and product specifications confirming the information.

Synergies:

Vegetated roofs may contribute to the following credits:

- SSc1, Ch2F2: Storm Water Design: Quantity Control
- SSc1, Ch2F3: Storm Water Design: Quality Control

Vegetated roofs may have a negative impact on the following credits:

- SSc1, Ch2F7: Water Efficient Landscaping: Reduced by 50%
- SSc1, Ch2F8: Water Efficient Landscaping: No Potable Water Use or No Irrigation
- SSc1, Ch2F10: Water Use Reduction: 30% Reduction

Possible Strategies and Technologies:

Choose a base building with a vegetated roof and/or a high albedo roof to reduce heat absorption.

Extra Credit (Exemplary Performance):

The project may earn extra points under SSc1, Ch2F12: Other Quantifiable Environmental

Performance if the tenant can show that 100% of the roof (excluding skylights, photovoltaic panels, and HVAC equipment) is comprised of a vegetated roof system.

Project Phase:
Pre-Design

LEED Submittal Phase:
Design

Related Code or Standard:
None

Responsible Parties:
The tenant and the real estate agent

SSc1, Ch2F6: Light Pollution Reduction (1 point)

Considerations and Benefits:

1) **Environmental:**
This feature aims at rewarding base buildings which reduce development impacts on nocturnal environments, reduce glare, improve nighttime visibility, and reduce sky-glow and light trespass from the site and building to increase night sky access.

   Reducing light pollution allows the observation of night sky, encourages nocturnal wildlife to thrive, provides improved visibility and visual comfort, and makes the building non intrusive to the community.

2) **Financial:** A well-designed lighting system can provide the proper amount of lighting at the right time and places. Choosing a base building that meets SSc1, Ch2F6 can substantially reduce the tenant's energy use and costs associated with lighting.

Submittals:

1) Drawings showing the locations of automatic controls, sequence of operation, and specifications or the building operation plan.
2) Drawings and specifications for automatic shading devices, or cutsheets (product data) showing that they block 90% of the light, sequence of operation, and specifications or the building operation plan.

Synergies:
Reducing light pollution may contribute to the following credit:

- EAc1.1: Optimize Energy Performance: Lighting Power
- EAc1.2: Optimize Energy Performance: Lighting Controls

It may also relate to the following:

- IEQc6.1: Controllability of Systems: Lighting

Possible Strategies and Technologies:

Choose a base building that reduces light pollution, and/or meets one or more of the following criteria:

1) Establish site lighting criteria to avoid night sky pollution and off-site lighting, but still maintain safe light levels.
2) Use a computer to model the site lighting and minimize site lighting when possible.
3) Use low-angle spotlights, low-reflectance surfaces, and full cutoff luminaires to reduce light pollution.

The local USGBC chapters or **Illuminating Engineering Society of North America (IESNA)** may have a list of the base buildings that meet the criteria of SSc1, Ch2F6.

Related information that you should know:

Per IESNA RP-33, all projects shall be classified in one of the following zones and shall comply with the related requirements:

LZ1-Dark (developed areas within rural areas, national parks, and state parks forest land. The population density is less than 200 people per square mile).
Design building exterior lights and site lights to produce an initial illuminance value no higher than 0.01 vertical and horizontal foot candles at the site boundary and beyond. Document that 0% of the total initially designed fixture lumens are emitted at an angle of 90 degree or higher from nadir (straight down).

LZ2-Low (light industry with limited nighttime use, neighborhood business district, primarily residential zones and residential mixed-use areas. The population density is between 200 and 3,000 people per square mile).
Design building and site mounted lights and exterior lights to produce an initial illuminance value no higher than 0.10 vertical and horizontal foot candles at the site boundary and no higher than 0.01 horizontal foot candles 10 feet beyond the site boundary. Document that 2% or less of the total initially designed fixture lumens are emitted at an angle of 90 degree or higher from nadir (straight down).

LZ3-Medium (All areas not included in LZ1, LZ2, or LZ3, like commercial/industrial, and high-density residential. The population density is more than 3,000 people per square mile).
Design building and site mounted lights and exterior lights to produce an initial illuminance value no higher than 0.20 vertical and horizontal foot candle at the site boundary and no higher than 0.01 horizontal foot candles 15 feet beyond the site boundary. Document that 5% or less of the total initially designed fixture lumens are emitted at an angle of 90 degree or higher from nadir (straight down).

LZ4-High (high activity commercial districts in major metropolitan areas, like major city centers and entertainment districts. The population density is more than 100,000 people per square mile).
Design building and site mounted lights and exterior lights to produce an initial illuminance value no higher than 0.60 vertical and horizontal foot candle at the site boundary and no higher than 0.01 horizontal foot candles 15 feet beyond the site boundary. Document that 10% or less of the total initially designed fixture lumens are emitted at an angle of 90 degree or higher from nadir (straight down).

For **LZ2, LZ3** and **LZ4**, if your project boundary is next to a public right-of-way, you can use the curb line instead of the site boundary for light trespass requirements.

For **all zones**, for a single luminaire and the related illuminance located at the intersection of a public roadway accessing the site and a private driveway, you can use the centerline of the public roadway as the site boundary for a length of twice the driveway width centered at the centerline of the driveway.

Extra Credit (Exemplary Performance):
None

Project Phase:
Schematic Design

LEED Submittal Phase:
Design

Related Code or Standard:
None

Responsible Parties:
The tenant and the real estate agent

SSc1, Ch2F7: Water Efficient Landscaping: Reduced by 50% (Outdoor Water, 2 points)

AND
SSc1, Ch2F8: Water Efficient Landscaping: No Potable Water Use or No Irrigation (Outdoor Water, 2 points in addition to Feature 7)

Considerations and Benefits:

1) **Environmental:**
   Landscape irrigation consumes 30% of the 26 billion gallons of potable water used daily in the U.S. Reducing the use of subsurface or surface natural water or potable water available on or near the job site for landscape irrigation can save a large amount of water or even eliminate the use of potable water.

   Adapted or native plants can save water, integrate well with natural surroundings, attract native wildlife, require less pesticides and fertilizer, and reduce environmental impacts, like degrading water quality, etc.

   Water efficient landscaping can save valuable potable water, and maintain natural aquifer.

2) **Financial:** Choosing a base building that meets SSc1, Ch2F7 and SSc1, Ch2F8 can reduce the tenant's water costs.

Submittals:

1) Estimations of baseline and design cases to show percentage of water use reduction
2) Description of which portion of the landscape irrigation will come from nonpotable sources
3) A landscape plan showing an irrigation system and a planting schedule. Contact the landlord to obtain a copy of the landscape plan if available.

Synergies:

Capturing rainwater may contribute to the following credits:

- SSc1, Ch2F2: Storm Water Design: Quantity Control
- SSc1, Ch2F3: Storm Water Design: Quality Control

Plantings can provide shade for hardscape and contribute to the following:

- SSc1, Ch2F4: Heat Island Effect: Non-Roof
- EAc1: Optimize Energy Performance

Vegetated roofs may contribute to the following:

- SSc1, Ch2F5: Heat Island Effect: Roof

Possible Strategies and Technologies:

Choose a base building with water efficient landscaping, and/or meet one or more of the following criteria:

1) Perform a climate and/or soils analysis and choose native or adapted plants to reduce irrigation water use.
2) Use graywater systems or rainwater harvest systems for landscape irrigation
3) Use a climate-based controller and/or high efficiency equipment, like a drip irrigation system or micro-irrigation system.
4) Use mulching, composting, and alternative mowing to minimize turf area.

A project with no landscape, OR a well constructed specifically to collect ground water for irrigation does not qualify for this point.

If a project has no landscape on the ground, but has outdoor planters, vegetated roofs, or courtyard garden space covering 5% or more of the site, it can earn points by reducing potable water use for irrigation for plants in these spaces or planters. If the area for outdoor planters, vegetated roofs, or courtyard garden space is less than 5% of the site, then the project can NOT earn any point.

Extra Credit (Exemplary Performance):

None

Project Phase:

Pre-Design

LEED Submittal Phase:

Design

Related Code or Standard:

None

Responsible Parties:

The tenant and the real estate agent

SSc1, Ch2F9: Innovative Wastewater Technologies (Indoor Water, 2 Points)

Considerations and Benefits:

1) **Environmental:**
   Toilets and urinals can use non-potable water, saving valuable water supplies. Reducing or eliminating sewage water can reduce chemical inputs, demands on infrastructure, the amount of water treated in treatment facilities, and related emissions and energy use. The project team should evaluate the pros and cons of on-site treatment vs. off-site treatment.

   On-site treatment changes "waste" into useful resources (potable and nonpotable water) and nutrients for plants, and creates an opportunity to educate the public on the LEED process and concept of sustainability.

2) **Financial:** Buildings that produce large amounts of waste water can save a tremendous amount of money by employing innovative wastewater technologies. High-efficiency urinals (HEU) and toilets (HET) are only slightly more costly than their conventional counterparts. Harvesting rainwater and recycling graywater may require a higher initial investment, and could affect leasing rates. Choosing a base building that meets SSc1, Ch2F9 can reduce the tenant's water costs.

Submittals:

1) Calculation of number and type of occupants
2) Manufacturer's data showing manufacturer, model, water consumption rate of each fitting and fixture
3) Information on system schematics and capacity of graywater and rainwater systems
4) A definition of user group
5) A list of plumbing fixtures based on user groups

Synergies:

Reusing graywater, harvesting rainwater, and reducing demand on local water aquifers may contribute to the following prerequisite and credits:

- SSc1, Ch2F2: Storm Water Design: Quantity Control
- SSc1, Ch2F3: Storm Water Design: Quality Control
- SSc1, Ch2F7: Water Efficient Landscaping: Reduced by 50%
- SSc1, Ch2F8: Water Efficient Landscaping: No Potable Water Use or No Irrigation
- SSc1, Ch2F10: Water Use Reduction: 30% Reduction
- WEp1: Water Use Reduction
- WEc1: Water Use Reduction

On the other hand, active systems for on-site wastewater treatment or reuse may require extra energy, and may affect the following prerequisite and credits:

- EAp1: Fundamental Commissioning of the Building Energy Systems

- EAc2: Enhanced Commissioning
- EAc3: Measurement and Verification

Possible Strategies and Technologies:

Choose a base building employs innovative wastewater technologies, and/or meets one or more of the following criteria:

1) Reuse graywater or storm water for mechanical and/or natural on-site waste treatment system, or sewage conveyance.
2) Use water-saving and waterless plumbing fixtures like waterless urinals and composting toilet systems to reduce wastewater volumes.
3) Use high-efficiency filtration systems, construct wetlands, or use a packaged biological nutrient removal system.

Check with governing agencies regarding codes and permits for on-site blackwater, graywater, or rainwater systems.

Extra Credit (Exemplary Performance):

None

Project Phase:

Pre-Design

LEED Submittal Phase:

Design

Related Code or Standard:

1) Energy Policy Act (EPAct) of 1992 (and as amended)
2) Energy Policy Act (EPAct) of 2005
3) International Associations of Plumbing and Mechanical Officials, Publication IAPMO/American National Standards Institute UPC 1-2006, Uniform Plumbing Code 2006, Section 402.0, Water-Conserving Fixtures and Fittings
4) International Code Council, International Plumbing Code 2006, Section 604, Design of Building Water Distribution System

Responsible Parties:

The tenant and the real estate agent

SSc1, Ch2F10: Water Use Reduction: 30% Reduction (1 point)

Considerations and Benefits:

1) **Environmental:**
   Same as WEc1, except WEc1 is for the whole building, while SSc1, Ch2F10 is for the tenant spaces only.

Submittals:

Same as SSc1, Ch2F9

Synergies:
Same as WEc1

Possible Strategies and Technologies:
Choose a base building meets the criteria of SSc1, Ch2F10 (30% reduction). Further, the building owner needs to have an ongoing plan requiring future tenants to comply with these regulations.

This feature is for LEED tenant projects that use 50% or less of the building: this is to prevent larger projects that use over 50% of the building from receiving double points for WEc1.

Extra Credit (Exemplary Performance):
The project may earn extra points under SSc1, Ch2F12: Other Quantifiable Environmental if the tenant can show the whole building has achieved 40% water use reduction.

Project Phase:
Pre-Design

LEED Submittal Phase:
Design

Related Code or Standard:
Same as SSc1, Ch2F9

Responsible Parties:
The tenant and the real estate agent

SSc1, Ch2F11: On-Site Renewable Energy (1 to 2 point)

Considerations and Benefits:

1) **Environmental:**
The process of energy production creates pollutants such as carbon dioxide, nitrogen oxide, and sulfur dioxide. These pollutants contribute to climate change, smog, and acid rain, and related negative impacts on humans, especially within the respiratory system.

Renewable energy provides great environmental benefits. For example, the use of biomass can help to remove 136 million tons of woody waste from landfills annually;

2) **Financial:**
Choosing a tenant space in a base building that uses on-site renewable energy can save the tenant's energy costs. **Net metering** (selling extra energy back to the utility), utility rebates, and long-term energy savings can offset the initial costs for renewable energy system. The projects team needs to confirm if these options are available for the project.

Climate, geography, available technologies, and other local factors can affect the decision about selecting the on-site renewable energy system.

Submittals:

1) Documentation for total annual energy generation, on-site renewable energy type, and backup energy source
2) Calculation of the amount of energy from each on-site renewable source
3) Documentation from the landlord confirming the capacity and performance of the on-site renewable energy systems; make sure renewable energy is not double-counted.
4) Keep documents for any incentives to support the on-site renewable energy systems installation.

Synergies:

On-site renewable energy equipment typically has a small impact on other credits. It requires measurement and verification, commissioning, and it also affects building energy performance. It can also reduce the amount of green power purchased. On-site renewable energy may contribute to the following credits:

- EAp1: Fundamental Commissioning of the Building Energy Systems
- EAp2: Minimum Energy Performance
- EAc1: Optimize Energy Performance
- EAc3: Measurement and Verification
- EAc4: Green Power

Possible Strategies and Technologies:

Choose a base building meets the criteria of SSc1, Ch2F11, and/or meet one or more of the following criteria:

1) Evaluate the project for biogas, biomass, or any other renewable and non-polluting energy strategies.
2) Use net metering to send extra on-site renewable energy back to the grid. Coordinate with the local utility company regarding potential rebate and incentive programs.
3) Use virtual energy (annual energy cost divided by annual energy consumption) and local utilities rates to assign a dollar value.
4) Geothermal heating systems, wind energy systems, wave and tidal systems, biofuel-base electrical systems, low-impact hydro electric systems, photovoltaic systems, and solar thermal systems are **eligible systems** for this credit; architectural features, daylighting and passive solar strategies, geo-exchange systems (ground source heat pumps) and purchased/green power (these should be part of EAc4) are **ineligible on-site systems** for this credit. Architectural features, daylighting and passive solar strategies, and geo-exchange are **ineligible** because they are already covered in EAp2 and may be considered under EAc1.
5) You can include the following biofuels as part of **renewable energy** for this credit: agricultural waste or crops, organic waste including animal waste, landfill gas, and mill residues and other untreated wood waste.
6) You cannot include the following biofuels as part of **renewable energy** for this credit: municipal solid waste combustion, wood coated with Formica, plastic or paint, preserved wood treated with materials containing arsenic, chlorine compounds, chromate copper arsenate, halide compounds, or halogens. If over 1% of the preserved wood fuel has been treated with these compounds, the energy system becomes ineligible.
7) Extra on-site renewable energy sold back to the grid cannot be included in the credit calculation.
8) The project team must retain the environmental attributes for on-site renewable energy: extra on-site renewable energy MUST be sold back at the market rate, and not at a higher premium. A higher premium means the environmental attributes have not been retained, and the project team cannot take credit for that energy as renewable.

9) **Exception:** GBCI will allow the sale of renewable energy certificates (RECs) from an on-site renewable energy system that claims credit if the energy system owner and/or the building owner meets the following:
   - The building owner purchases Green-e eligible RECs that are equivalent to 2 times the on-site system's annual rated energy output from another source.
10) Most in-state RECs are more expensive, and they are different from out-of-state RECs for two reasons: a. Both RECs have the same CO_2 reduction, but out-of-state RECs have fewer other emission reductions; b. Out-of-state RECs may be limited by the design and technical potentials.
   - In-state RECs can be replaced by out-of-state RECs at a ratio of 1:2.
   - The seller of the on-site RECs must follow all related guidelines and not claims the system's environmental attributes.

Extra Credit (Exemplary Performance):
The project may earn extra points under SSc1, Ch2F12: Other Quantifiable Environmental Performance if the on-site renewable energy equals 10% of the building's energy cost.

Project Phase:
Pre-Design

LEED Submittal Phase:
Design

Related Code or Standard:
1) ANSI/ASHRAE/IESNA Standard 90.1-2007, Energy Standard for Buildings Except Low-Rise Residential (with errata but without addenda)
2) IESNA
3) Department of Energy (DOE)
4) Commercial Building Energy Consumption Survey (CBECS)
5) EIA (Energy Information Administration)

Responsible Parties:
The tenant and the real estate agent

SSc1, Ch2F12: Other Quantifiable Environmental Performance (1 point)

Other Quantifiable Environmental Performance means:
1) Green feature based on the criteria another LEED rating system credit

OR

2) The exemplary performance of any of the **feature** of the credit.

   The green feature based on the criteria another LEED rating system credit must be different from the ones for LEED CI.

Considerations and Benefits:
1) **Environmental:**
   SSc1, Ch2F12 rewards projects for green building features in other LEED rating systems not

covered in SSc1, Ch2F1 through SSc1, Ch2F11. For example, for LEED NC, CS, and Schools rating systems, SSc5.1: Site Development: Protect or Restore Habitat, the project can earn one innovation point by protecting or restoring at least 75% instead of 50% (or ½) of the site area (not including the building footprint) with adapted or native vegetation on a previously graded or developed site.

SSc1, Ch2F12 also rewards projects for green building features covered in the **exemplary performance** portion of SSc1, Ch2F1 through SSc1, Ch2F11.

Submittals:

1) Documentation (calculations, a brief narrative, etc.) for the environmental benefits or exceptional performance achieved through green building features in other LEED rating systems not covered in SSc1, Ch2F1 through SSc1, Ch2F11

   OR
2) Innovative performance in green building categories covered in the **exemplary performance** portion of SSc1, Ch2F1 through SSc1, Ch2F11.

Synergies:

1) See related credits section in other LEED rating systems not covered in SSc1, Ch2F1 through SSc1, Ch2F11

   OR
2) Innovative performance in green building categories covered in the **exemplary performance** portion of SSc1, Ch2F1 through SSc1, Ch2F11.

Possible Strategies and Technologies:

1) Choose a base building that has a minimum of 1 credit achieved through green building features in other LEED rating systems not covered in SSc1, Ch2F1 through SSc1, Ch2F11

   OR
2) Other ID points not covered by any LEED rating systems. Submit a CIR unless you can find a precedent in a previous LEED CI project (1 point for each credit that offer 1 extra point covered in both paths above)

   OR
3) Innovative performance in green building categories covered in the **exemplary performance** portion of SSc1, Ch2F1 through SSc1, Ch2F11. (A maximum of 1 extra point)

Extra Credit (Exemplary Performance):

Projects earning exemplary performance under SSc1 cannot earn additional exemplary performance under ID section.

Project Phase:

Pre-Design

LEED Submittal Phase:

Design

Related Code or Standard:

See related credits

Responsible Parties:

The tenant and the real estate agent

SSc2: Development Density and Community Connectivity (6 Points)

Purpose:

1) To protect natural resources, habitat, and greenfields.
2) Encourage development in urban areas with existing infrastructure.

Credit Paths:

1st Path: Development Density

Choose a space in a building on a walkable, developed site AND in an area with at least 60,000 s.f. per acre net density. You must include the proposed project area and base the density calculation on a two-story downtown development.

OR

2nd Path: Community Connectivity

Choose a space in a building on a developed site AND within ½ mile of at least 10 Basic Service AND within ½ mile of a neighborhood or residential area with at least 10 units per acre net density AND with pedestrian access between the service and the building. Note: The ½ mile radius can be drawn from a main building entrance and can include the services within the radius. For mixed-use project, you can only count 1 service in the project boundary, and it must be open to the public.

The qualified Basic Services include, but are not limited to:

a) Place of Worship;
b) Restaurant;
c) Supermarket;
d) Convenience Grocery;
e) Laundry;
f) Cleaner;
g) Beauty Salon;
h) Hardware;
i) Pharmacy;
j) Medical/Dental;
k) Bank;
l) Senior Care Facility;
m) Community Center;
n) Fitness Center;
o) Daycare;
p) School;
q) Library;
r) Museum;

s) Theater;
t) Park;
u) Fire Station;
v) Post Office.

Up to 2 of the 10 basic services can be anticipated.

You can only count one of each service, except two restaurants may be counted.

Projects not using existing infrastructure and greenfield developments do not qualify for this credit.

Considerations and Benefits:

1) **Environmental:** Choosing a base building meets SSc2 requirements reduces green house emissions, air pollution, and other traffic impacts.

   Rehabilitating existing buildings in cities is a way to reduce demand for new materials. The potential disadvantages for this approach are: undesirable air quality, contaminated soils, limited open space and daylight opportunities, and lower IEQ, etc.

2) **Financial:** Choosing a base building that meets SSc2 requirements fully takes advantage of existing infrastructure (utilities, roads, etc.), controls urban sprawl, and improves the local economy. A base building at a site close to mass transit may substantially reduce the number of parking spaces needed and the related costs. A base building located at a walkable distance to basic services can increase the level of physical activities and reduce driving time for the building occupants, which in turn will improve their health and productivity and save the tenant money.

Submittals:

For 1st Path

1) A site vicinity plan highlighting the development density and its radius
2) Records of building and site area

For 2nd Path

1) A site vicinity plan highlighting locations of the residential areas, the ½ mile radius, and a list highlighting the location and category of qualifying services and residential areas.

Synergies:

Channeling development to previously developed urban areas near public transportation may contribute to:

- SSc1: Site Selection
- SSc3.1: Alternative Transportation: Public Transportation Access

Possible Strategies and Technologies:

Select urban sites that comply with credit paths 1 and 2 above.

Extra Credit (Exemplary Performance):

You can get one innovation point by doubling the requirements, i.e. 120,000 s.f per acre.

Project Phase:
Pre-Design

LEED Submittal Phase:
Design

Related Code or Standard:
None

Responsible Parties:
The tenant and the real estate agent

SSc3.1: Alternative Transportation: Public Transportation Access (6 Points)

Purpose:
To reduce automobile use, land development impact, and pollution.

Credit Paths:
1) The project shall be located within ½ mile of one planned and funded, or one existing commuter rail or subway station.

OR
2) The project shall be located within ¼ mile of at least one stop for two or more campus or public bus lines usable by the occupants of the tenant space.

Note: All distances above shall be measured from a main building entrance. You can only count a school's bus system as one bus line.

For both paths:
Provide dedicated bike lanes and walking paths to transit lines.

Considerations and Benefits:
1) **Environmental:** Solo drivers contribute to air pollution and other environmental problems. People will use alternate transportation means if they are convenient. **Mass transit** reduces energy use and the related greenhouse gas emissions, and demand for parking spaces, which in turn leave spaces for greater development density or natural areas.

   Public transportation uses ½ of the fuel when compared to private vehicles, and can save 45 million barrels of oil each year in the U.S. It also reduces the demand for infrastructure (impervious roads and parking facilities, etc).

2) **Financial:** Proximity to mass transit is a benefit, and can increase a building's value. It can help the tenant attract employees. Reducing the size of the parking area will lower **Common Area Maintenance (CAM)** costs, reduce the number of impervious areas, and lower the related storm

water fees.

Submittals:

1) Identify bus routes and rail stations serving the project
2) Create a site vicinity map (to scale) with the walking paths between bus stops, rail stations, and the project building's main entrance labeled

Synergies:

Projects close to public transportation tend to be in previously developed and densely developed areas. They may qualify for:

- SSc1: Site Selection
- SSc2: Development Density and Community Connectivity

Possible Strategies and Technologies:

Choose a base building that meets one or more of the following criteria:

1) Locate the building close to mass transit.
2) Survey future building occupants to determine their transportation needs.
3) Create a transportation management plan.

Extra Credit (Exemplary Performance):

You can receive one innovation point by:

1) Create a transportation management plan to show a quantifiable reduction in the number of solo drivers via various transportation alternatives (1 point maximum).

OR

2) Doubling the requirements, i.e., two train stations or four bus lines.

AND

3) These stops must have a minimum of 200 transit rides (service frequency) in total per day.

Project Phase:

Pre-Design

LEED Submittal Phase:

Design

Related Code or Standard:

None

Responsible Party:

The tenant and the real estate agent

SSc3.2: Alternative Transportation: Bicycle Storage and Changing Rooms (2 points)

Purpose:

1) To reduce automobile use.
2) To reduce land development impact and pollution.

Credit Paths:

1) Within 200 yards of the building entrance, provide secure bicycle storage/rack for at least 5% of the tenant occupants, AND provide changing rooms and showers for 0.5% of **Full-Time Equivalent (FTE)** users within 200 yards of the entrance for the building, or in the building. Note: FTE Occupants=Occupants Hours/8

   See Appendixes for "Default Occupancy Factors" Table.

Considerations and Benefits:

1) **Environmental:** Automobiles contribute to smog, air pollution, and other environmental problems. **Bicycling** reduces energy use and the related greenhouse gas emission, and demand for parking spaces, which in turn leave spaces for greater development density or natural areas. It also alleviate noise pollution, traffic congestion, and require very little infrastructure (impervious roads and parking facilities, etc).

   Bicycles can replace automobiles for short trips. This can be very beneficial since most vehicle emissions occur in the first few minutes of driving.

2) **Financial:** The initial cost for bicycle storage and changing rooms is very low. If a building provides these facilities and encourage bicycling, it can be very beneficial to the occupants' health. **Bicycling and walking** make people healthier, and create additional opportunities for interaction and enjoyment of the surrounding area.

Submittals:

1) Calculate the number of occupants for each use type and determine the number of showering facilities and bicycle storage units needed
2) Create a plan showing the quantity and locations of showering and bicycle storage facilities, also determine the distance between the project building's entry and the facilities

Synergies:

Paving materials for on-site bicycle lanes may affect the following credits:

- SSc1, Ch2F2: Storm Water Design: Quantity Control
- SSc1, Ch2F3: Storm Water Design: Quality Control
- SSc1, Ch2F4: Heat Island Effect-Non-Roof

Possible Strategies and Technologies:

Choose a base building with:

1) Permanent shower/changing rooms and/or
2) Permanent bicycle storage/racks: If they are not for the tenant's exclusive use, they should be

based on the population of the whole building.
3) Convenient access to safe bicycle pathways.

Extra Credit (Exemplary Performance):
You may get only one extra point for all credits under SSc3 by implementing a comprehensive program that results in quantifiable automobile use reduction.

Project Phase:
Schematic Design

LEED Submittal Phase:
Design

Related Code or Standard:
None

Responsible Party:
The tenant and the real estate agent

SSc3.3: Alternative Transportation: Parking Availability (2 points)

Purpose:
1) To reduce automobile use.
2) To reduce land development impacts and pollution.

Credit Paths:
For Projects Occupying Less Than 75% of the Total Building Area:
1) Provide parking spaces for tenant to meet, but not to exceed, minimum local zoning requirements.

   AND
   Provide preferred parking (based on the total parking capacity of the site) for vanpools or carpools for at least 5% of tenant occupants.

   OR
2) Do not subsidize or provide parking for tenant occupants

For Projects Occupying 75% or More of the Total Building Area:
1) Provide parking capacity to meet, but not to exceed, minimum local zoning requirements

   AND
   Provide preferred parking (based on the total parking capacity of the site) for vanpools or carpools for at least 5% of building occupants.

   OR

2) Do not add new parking for rehabilitation jobs.

   AND
   Provide preferred parking (based on the total parking capacity of the site) for vanpools or carpools for at least 5% of building occupants.

Considerations and Benefits:

1) **Environmental:** Automobiles contribute to smog, air pollution, and other environmental problems. **Carpooling** reduces energy use and the related greenhouse gas emission, and demand for parking spaces, which in turn leave spaces for greater development density or natural areas. It also alleviate noise pollution, traffic congestion, and require very little infrastructure (impervious roads and parking facilities, etc).

2) **Financial:** Carpooling reduces the demand for parking spaces, infrastructure, as well as the related costs. Smaller parking areas also reduce impervious surfaces, and the related storm water runoff and treatment costs. Many governing agencies also have tax incentives for carpooling.

Submittals:

1) Information on:
   - The type and amount of parking
   - Programming and/or infrastructure supporting vanpool and carpool initiatives like the number of FTEs and preferred parking spaces for the tenant
   - Zoning requirements
2) A lease agreement or brochure for occupants explaining the vanpooling and carpooling support structure
3) A site plan indicating parking spaces, the number and locations of preferred vanpool or carpool spaces for the tenant

Synergies:

Meeting but NOT exceeding the zoning requirements, and minimizing parking, especially surface parking can contribute to the following credits:

- SSc1, Ch2F2: Storm Water Design: Quantity Control
- SSc1, Ch2F3: Storm Water Design: Quality Control
- SSc1, Ch2F4: Heat Island Effect-Non-Roof

Possible Strategies and Technologies:

Choose a base building that is LEED certified and has achieved this credit, or a building that has met one or more of the following criteria:

1) Does not **guarantee** more spaces for the tenant than required by codes. The **guarantee** of more spaces including exclusive parking spaces, and number of access cards for restricted parking areas assigned for the tenant and his guests.
2) Encourages shared vehicle use.
3) Shares parking with neighboring buildings.
4) Minimizes parking lot size.

Related information for base building:

Only vehicles classified as **Zero Emission Vehicles (ZEV)** by California Air Resources Board or vehicles with a green score of at least 40 on the **American Council for an Energy Efficient Economy (ACEEE)** annual vehicle rating guide are qualified as fuel efficient and low emitting vehicles for LEED credit of a base building.

Extra Credit (Exemplary Performance):
You may get only one extra point for all credits under SSc3 by implementing a comprehensive program that results in quantifiable automobile use reduction.

Project Phase:
Schematic Design

LEED Submittal Phase:
Design

Related Code or Standard:
None

Responsible Party:
Architect and Owner

SS Summary and Mnemonics:

SS Credit Name	Extra Credit	Responsible Party
SSc1: **S**ite Selection	See SSc1, Ch2F12 below	The tenant and the real estate agent
Ch1 (Choice One): Choose a building that is **L**EED Certified (5 points).	0	The tenant and the real estate agent
SSc1, Ch2F1: **B**rownfield Redevelopment (1 point)	0	The tenant and the real estate agent
SSc1, Ch2F2: **S**torm Water Design: Quantity Control (1 point)	0	The tenant and the real estate agent
SSc1, Ch2F3: Storm Water Design: Quality Control (1 point)	0	The tenant and the real estate agent
SSc1, Ch2F4: **H**eat Island Effect: Non-Roof (1 point)	The project may earn extra points under SSc1, Ch2F12: Other Quantifiable Environmental Performance if the tenant can show the building has met 2 or more of the credit paths in SSc1, Ch2F4.	The tenant and the real estate agent
SSc1, Ch2F5: Heat Island Effect: Roof (1 point)	The project may earn extra points under SSc1, Ch2F12: Other Quantifiable Environmental Performance if the tenant can show 100% of the roof (excluding skylights, photovoltaic panel and HVAC equipment) is made of a vegetated roof system.	The tenant and the real estate agent
SSc1, Ch2F6: **L**ight Pollution Reduction (1 point)	0	The tenant and the real estate agent
SSc1, Ch2F7: Water Efficient **L**andscaping: Reduced by 50% (Outdoor Water, 2 points) AND SSc1, Ch2F8: Water Efficient Landscaping: No Potable Water Use or No Irrigation (Outdoor Water, 2 points in addition to Feature 7)	0	The tenant and the real estate agent
SSc1, Ch2F9: **I**nnovative Wastewater Technologies (Indoor Water, 2 Points)	0	The tenant and the real estate agent
SSc1, Ch2F10: Water **U**se Reduction:30% Reduction (1 point)	0	The tenant and the real estate agent

SSc1, Ch2F11: **<u>O</u>**n-Site Renewable Energy (<u>1</u> to <u>2</u> point)	The project may earn extra points under SSc1, Ch2F12: Other Quantifiable Environmental Performance if the on-site renewable energy equals to 10% of the building energy cost.	The tenant and the real estate agent
SSc1, Ch2F12: Other **<u>Q</u>**uantifiable Environmental Performance (<u>1</u> point)	Projects earning exemplary performance under SSc1 can<u>not</u> earn additional exemplary performance under ID section.	The tenant and the real estate agent
SSc2: **<u>D</u>**evelopment Density and Community Connectivity (<u>6</u> Points)	0	The tenant and the real estate agent
SSc3.1: **<u>A</u>**lternative Transportation: Public Transportation Access (<u>6</u> Points)	You can get <u>one</u> innovation point by: 1) Create a transportation management <u>plan</u> to show quantifiable reduction of the solo drivers via various transportation alternatives (1 point maximum). OR 2) <u>Doubling</u> the requirements, i.e., <u>two</u> train stations or <u>four</u> bus lines. AND 3) These stops must have a minimum of 200 transit rides (service frequency) in total per day.	The tenant and the real estate agent
SSc3.2: Alternative Transportation: Bicycle Storage and Changing Rooms (<u>2</u> points)	You may get only <u>one</u> extra point for all credits under SSc3 by implementing a comprehensive program that results in <u>quantifiable</u> automobile use reduction.	The tenant and the real estate agent
SSc3.3: Alternative Transportation: Parking Availability (<u>2</u> points)	Same as above	<u>Architect</u> and Owner

Mnemonics: **<u>S</u>**andy **<u>L</u>**oves **<u>B</u>**rad **<u>S</u>**mith. **<u>H</u>**is **<u>L</u>**ove **<u>L</u>**ife **<u>I</u>**s **<u>O</u>**ften **<u>Q</u>**uite **<u>D</u>**angerous and **<u>A</u>**dventurous. (See bold and underlined letters in the credit names on table above also). Of course, you can create your own **mnemonics** to help you memorize the credits above as well.

B. Water Efficiency (WE):

Overall purpose:
Water Use Reduction (Indoor Water)

Core concepts:
1) Regulation of indoor water
   - Save as much indoor potable water as possible
   - Use water efficiently

Recognition, regulation and incentives:
1) Recognition
   - WaterSense product label sponsored by the EPA
2) Regulation (Requirements and Goals)
   - Mandatory federal water efficiency requirements per the Energy Policy Act (EPAct) of 2005 and Executive Order 13423 (2007): All federal facilities shall reduce water use intensity by 2% per year between 2006 and 2015 to reach a total reduction of 20%.
   - Energy Policy Act (1992): Mandatory requirements requiring the use of water conserving plumbing fixtures in industrial, commercial, and residential buildings.
3) Financial incentives
   - Local/state rebates and credits for water saving devices

Overall strategies and technologies:
1) **Reduce indoor potable water demand**
   Use non-potable water: reuse graywater, capture and use rainwater, reduce water use via innovative wastewater treatment, use water efficiently: use waterless or high efficiency fixtures.

Specific Technical Information:

WEp1: Water Use Reduction (Indoor Water, Mandatory, 0 points)

Purpose:
To reduce the pressure on wastewater systems and municipal water supplies by saving as much water as possible within buildings.

Note: This prerequisite is for reducing indoor water use. Some questions on practice exams or real LEED exams may test your understanding of whether a prerequisite or a credit is related to indoor water use or outdoor water use. I have read these types of sample questions on LEED exams.

Credit Path:
Meet the Energy Policy Act of 1992 fixture performance requirements, and then apply strategies to

save 20% of water when compared with the water use baseline calculation for the building (not including landscape irrigation). The calculation should only include urinals, water closets, showers, pre-rinse spray valves, kitchen sink faucets, lavatory faucets and shall be based on estimated occupant use.

Some buildings have several shifts, and may include the following:
1) Residents
2) Full-time staff
3) Part-time staff
4) Transients (retail customers, student visitors, etc.)

The number of occupants can be identified by the **occupancy type**, i.e., two people for a one-bedroom unit, and three people for a two-bedroom unit, etc. If you do not know the occupancy (like CS projects or mixed-use projects in the early design stages where you do not know who will be the tenants), then you can use the "Default Occupancy Factors" in the appendixes. This table is included in the book for your convenience.

FTE = number of hours of occupancy/8

You can estimate the number of transients for your projects. Use a daily average number over a one year period.
Use your best judgment to decide if someone should be reported as a FTE or a transient. For example, a volunteer who works at the school for four hours each day can be considered a FTE with a value of 0.5, and an individual who attends a basketball game can be reported as a visitor.
Use a 1 to 1 male to female ratio for your projects unless they have a specific ratio. Any special ratios must be described with a narrative.

Important data to memorize: (The numbers in the two following tables are very important. Almost every LEED exam tests them)

Fixture uses per day:

Non-residential Projects					
	Water Closet	Urinal	Lavatory (15 sec, 12 sec with autocontrol)	Shower (300 sec)	Kitchen Sink (15 sec)
FTE (including Student FTE)					
Female	3	0	3	0.1 0 for student FTE	1
Male	1	2	3	0.1 0 for student FTE	1
Transients (Student/Visitors)					
Female	0.5	0	0.5	0	0
Male	0.1	0.4	0.5	0	0
Transients (Retail Customers)					
Female	0.2	0	0.2	0	0
Male	0.1	0.1	0.2	0	0
Residential Projects					
	Water Closet	Urinal	Lavatory (60 sec)	Shower (480 sec)	Kitchen Sink (60 sec)
Female	5	N/A	5	1	4
Male	5	N/A	5	1	4

Note:

1) Lavatory faucets are counted as a 60 second duration per use for **residential** projects, a 15 second duration for **non-residential** projects, and a 12 second duration when equipped with an autocontrol.
2) Kitchen sink faucets are counted as a 60 second duration per use for **residential** projects, and a 15 second duration for **non-residential** projects
3) Showers are counted as a 480 second duration for **residential** projects, and a 300 second duration for **non-residential** projects.
4) This table applies to CI, NC, CS, Schools, as well as Commercial and Residential projects. I have combined several tables into one table to save you time.

Flow Rates (GPF or GPM or GPC):

	Water Closet	Urinal	Lavatory	Shower	Kitchen Sink
Conventional (baseline)	1.6 gpf except blow-out fixture at 3.5 gpf	1.0 gpf	**Residential** 2.2gpm at 60psi **Commercial** 2.2gpm at 60 psi for private app. 0.5gpm at 60 psi for public use 0.25gpc (gallons per cycle) for metering faucets	2.5 gpm at 80psi per shower stall	2.2gpm at 60 psi
IPC & UPC Standards	**Ibid** (Same as above)	Ibid	Ibid	Ibid	Ibid
EPA WaterSense Standard or High-efficiency	1.28 gpf	0.5 gpf	1.5 gpm for private lavatory faucets and aerators	1.5-2.0 gpm	
HET, single flush pressure assist	1 gpf	0.5 gpf for HEU			
HET, dual flush (full-flush)	1.6 gpf				
HET, dual flush (low-flush)	1.1 gpf				
HET, foam flush	0.05 gpf				
Low-Flow	1.1 gpf	0.5 gpf	1.8 gpm	1.8 gpm	1.8 gpm
Ultra Low- Flow	0.8 gpm		0.5 gpm		
Non-Water Urinal or Composting Toilet		0			

Note:

1) Private app. (application) means hospital patient rooms, motel, and hotel rooms, etc.
2) Commercial **pre-rinse spray valves** for food service application ≤ **1.6gpm**
3) This table applies to **non-residential** (NC, CS, school, commercial, etc.) projects and **residential** projects. I have combined several tables into one table to save you time.
4) This table is based on the **Energy Policy Act of 1992**, **EPA WaterSense** Standard, Uniform Plumbing Code (**UPC**), and International Plumbing Code (**IPC**) standards for plumbing fixture water use. **WaterSense** is a partnership program sponsored by the **EPA**. The **EPA WaterSense** standard exceeds the IPC and UPC requirements in some cases.
5) The following fixtures shall not be included in the water use calculations for this prerequisite, but they may be included for extra ID points for WRc3 for LEED NC, CS, and Schools (NOT LEED

CI), Water Use Reduction:
Automatic commercial ice makers, commercial dishwashers, commercial steam cookers, residential clothes washers, commercial (family-size) clothes washers, standard and compact residential dishwashers.

6) **High-efficiency (HE) fixtures** include: single-flush gravity fed toilets, high-efficiency toilets (HET), and high-efficiency urinals (HEU), etc.

   Note: HETs are available in different flush types: single-flush gravity fed (like a conventional toilet), single flush pressure assist, and dual-flush in both gravity fed and pressure assist models.

Considerations and Benefits:

1) **Environmental:** Reducing the potable water use within buildings can reduce our demand for water drawn from underground aquifers, streams, rivers, lakes, and other water bodies. It also saves valuable water for future generations, protects natural water cycles, and reduces costs for building and maintaining wastewater treatment and water supply infrastructure.

   It reduces energy use and the related greenhouse gas, reduces chemical input for water treatment.

2) **Financial:** Reducing the potable water use within buildings saves building operating costs, reduces the pressure on wastewater systems and municipal water supplies and the related infrastructure costs.

   Project team needs to evaluate the cost of water saving devices vs. actual savings. For example, tamper-proof faucet aerators on existing fixtures can be very cost-effective, while graywater recycling or rainwater-harvesting systems may require a larger investment. High-efficiency and waterless fixtures may have a higher initial cost, but may save more money in the long run. Their **ROI** (Return on Investment) may be 5 to 8 years. Some municipalities may give out free high-efficiency and waterless fixtures to their residents to encourage water savings.

Submittals:

1) Number and type of occupants
2) Manufacturer's data (cutsheets) showing manufacturer name, model, and water consumption rates of each fitting and fixture
3) A definition of the usage groups
4) A list of plumbing fixtures by usage group

Synergies:

Increasing the use of rainwater, graywater, and reducing the demand on local water aquifers may contribute to the following credits:

- SSc1, Ch2F2: Storm Water Design: Quantity Control
- SSc1, Ch2F3: Storm Water Design: Quality Control
- SSc1, Ch2F7 & F8: Water Efficient Landscaping
- SSc1, Ch2F9: Innovative Wastewater Technologies
- SSc1, Ch2F10: Water Use Reduction
- WEc1: Water Use Reduction

Water reuse may need extra energy, and active systems will need commissioning if part of tenant's work. This may affect the following:

- EAp1: Fundamental Commissioning of the Building Energy Systems
- EAc3: Enhanced Commissioning
- EAc5: Measurement and Verification

Possible Strategies and Technologies:

1) Use graywater or storm water for custodial uses, flushing urinals and toilets, and other non-potable use.
2) Use water-saving and waterless plumbing fixtures like waterless urinals and composting toilet systems, and occupant sensors to reduce wastewater volumes.
3) Consider water saving devices in lease negotiation.

Extra Credit (Exemplary Performance):

None

Project Phase:

Pre-Design

LEED Submittal Phase:

Design

Related Code or Standard:

1) Energy Policy Act (EPAct) of 1992 (and as amended)
2) Energy Policy Act (EPAct) of 2005
3) International Associations of Plumbing and Mechanical Officials, Publication IAPMO/American National Standards Institute UPC 1-2006, Uniform Plumbing Code 2006, Section 402.0, Water-Conserving Fixtures and Fittings
4) International Code Council (ICC), International Plumbing Code (IPC) 2006, Section 604, Design of Building Water Distribution System

Responsible Party:

MEP Engineer, Architect, and Owner

WEc1: Water Use Reduction (Indoor Water, 6 to 11 Points)

Purpose:

To reduce pressure on wastewater systems and municipal water supplies by saving more water within buildings.

Credit Path:

Meet the Energy Policy Act of 1992 fixture performance requirements, and then apply strategies to save water when compared with the water use baseline calculation for the building. Your calculation should only include urinals, water closets, showers, pre-rinse spray valves, kitchen sink faucets, and lavatory faucets and shall be based on estimated occupant use.

Percentage of Water Use Reduction	Points
30%	6
35%	8
40%	11

Note: This percentage of water use reduction is based on the water use baseline. See Tables and notes in **WEp1** for water use baseline information.

Considerations and Benefits:
See WEp1

Submittals:
See WEp1

Synergies:
See WEp1

Possible Strategies and Technologies:
1) Use graywater or storm water for custodial uses, flushing urinals and toilets, and other non-potable use.
2) Use water-saving and waterless plumbing fixtures like waterless urinals and composting toilet systems, and occupant sensors to reduce wastewater volumes.

Extra Credit (Exemplary Performance):
You can receive one innovation point by saving 45% of potable water.

Project Phase:
Schematic Design

LEED Submittal Phase:
Design

Related Code or Standard:
Energy Policy Act of 1992

Responsible Party:
MEP Engineer, Architect, and Owner

WE Summary and Mnemonics:

WE Credit Name	Extra Credit	Responsible Party
WEp1: Water Use Reduction (Indoor Water, Mandatory, 0 points)	0	MEP Engineer, Architect, and Owner
WEc1: Water Use Reduction (Indoor Water, 6 to 11 Points)	1 extra ID point by saving 45% of potable water.	MEP Engineer, Architect, and Owner

C. Energy and Atmosphere (EA)

Overall purpose:

1) Save energy
2) Promote the supply of renewable energy
3) Fundamental Commissioning of the Building Energy Systems
4) Minimum Energy Performance
5) Fundamental Refrigerant Management
6) Optimize Energy Performance
7) On-Site Renewable Energy
8) Enhanced Commissioning
9) Enhanced Refrigerant Management
10) Measurement and Verification
11) Green Power

Mnemonics:
So, Please Forward My Files!
Oh! Oprah Enters Energy Management Group. (See underlined letters above also).

Core concepts:

1) Energy Efficiency and Demand
   - Understand energy criteria
   - Save energy
   - Evaluate energy performance
2) The Supply of Energy
   - Buy off-site renewable energy
   - Generate on-site renewable energy

Recognition, regulation, and incentives:

1) **Recognition:** Energy Star Program, Target Finder Rating Tool, etc.
2) **Regulation (Requirements and Goals):** See detailed discussion under each credit.
3) **Incentives**
   - Private Sector: lower risk and lower premiums for property insurance, social responsibility of corporations, and availability of money
   - Public Sector: tax rebates and credits, incentive for development, expedited plan review and permit processing, and technology-based measures.

Overall strategies and technologies:

Note: Not **all** strategies and technologies have to be used simultaneously for your project.

1) Utilize typical energy use patterns for various building types
2) Use statistical databases: performance-based vs. prescriptive approaches
3) Use code-based energy models

Typical Energy Use Pattern Matrix

Building Type	Median Electrical Intensity (kWh/sf-yr)
Education	6.6
Office	11.7
Retail (except mall)	8.0
Food Sales	58.9
Food Service	28.7
Lodging	12.6

4) Reduce energy use: reduce the use of artificial HVAC and lighting; use energy-efficient equipment with feedbacks and controls
5) Consider building orientation and improve building envelope performance
6) Use Energy Star appliances and energy-efficient equipment
7) Measure and verify building energy performance: build and operate as designed (commissioning, as well as continuous and retro-commissioning); monitor performance and improvement over time (monitoring and verification)
8) Generate on-site renewable energy: geothermal, solar PV, and wind energy
9) Take advantage of the site: passive solar energy, natural ventilation, and passive cooling
10) Buy off-site renewable energy

Specific Technical Information:

EAp1: Fundamental Commissioning of the Building Energy Systems (Mandatory, 0 points)

Purpose:

To confirm that the building's energy systems are built and adjusted correctly, and that they will perform per the basis of design (BOD), construction documents, and owner's project requirements (OPR).

The **advantages (or benefits) of commissioning** are confirmation of a system's performance per the owner's requirements, fewer contractor callbacks, lower operating costs, less energy use, better building documentation, and increased occupant productivity.

Credit Path (0 point, mandatory requirements/prerequisites):

Your commissioning team shall complete the following commissioning plan/process per the related USGBC reference guide:

a. Assign a person to be the Commission Authority (CxA) to review, lead, and supervise the completion of the commissioning process.
   1) The CxA shall report the recommendation, findings, and results directly to the owner.
   2) The CxA shall have commissioning authority experience on at least 2 building projects.
   3) The CxA shall be independent of the project's design and construction management teams,

but he or she can be an employee of the firm providing the services. The CxA could be a qualified consultant or employee of the owner.
   4) The CxA can be a qualified member with the required experience on the project's design and construction teams for projects with less than 50,000 gross s.f.
b. Create and implement a commissioning plan.
c. Create and implement commissioning requirements into the construction documents.
d. Confirm the performance and installation of the system.
e. Finish a summary commissioning report. The report shall include:
   1) Executive summary
   2) Test results and evaluation
   3) History of deficiencies
   4) Confirmation from the CxA whether the systems meet the Basis of Design (BOD), Owner's Project Requirements (OPR), and Constructions Documents (CD)

If you are pursuing credit for EAc2, Enhanced Commissioning, also include:
   5) Submittal process summary
   6) Design review summary
   7) Operations and maintenance documentation, and training process
   8) As-built drawings

The design team needs to create the Basis of Design (BOD). The owner needs to document the Owner's Project Requirements (OPR). The CxA needs to review these for clarity and completeness. The design team and owner are responsible for updating their own documents.

This credit requires completion of the commissioning process activities for at least the following systems:
a. Wind, solar, or other renewable energy systems.
b. Daylighting and lighting controls (Electrical).
c. Mechanical and passive HVAC and refrigeration systems and related controls.
d. Domestic hot water systems (Plumbing).

Mnemonics: REMP (See underlined letters above)

Considerations and Benefits:

1) **Environmental:** If building energy systems do not perform as designed, they may consume substantially more energy in the long term. Commissioning can confirm a system's performance per the owner's requirements, reduce contractor callbacks, lower operating costs, reduce energy usage, result in improved building documentation, and increase occupant productivity.

2) **Financial:** If the project team has not included commissioning as part of the budget from the very beginning, owners/ tenants may oppose it because of the cost. However, if long-term benefits are included, commissioning is very cost-effective. It can reduce energy costs, increase occupant productivity, reduce tenant vacancy and turn-over, liabilities related to indoor air quality (IAQ), and prevent premature replacement of equipment.

Submittals:

1) Updated commissioning plan at strategic milestones of the project, including but not limited to the DD (Design Development) phase, the CD (Construction Documents) phase, and right before the kick-off meeting with the GC (general contractor).
2) A systems list with commissioned systems noted

3) Confirmation that CxA has documented experience on a minimum of 2 building projects
4) Copies of OPR, BOD, commissioning specifications and report, and system manual

Synergies:

A project team can include the following prerequisites and credits in fundamental commissioning:

a. Wind, solar or, other renewable energy systems:

1) SSc1, Ch2F11: On-Site Renewable Energy: Check performance.

b. Daylighting and lighting controls (Electrical):

2) SSc1, Ch2F6: Light Pollution Reduction: Confirm existing site, project and building lighting design comply with OPR.

3) EAp2: Minimum Energy Performance: Confirm the project has met the mandatory provisions and prescriptive requirements of ASHRAE 90.1-2007.
4) EAc1.1: Optimize Energy Performance: Lighting Power: Documentation of ASHRAE 90.1-2007 compliance.
5) EAc1.2: Optimize Energy Performance: Lighting Controls: Make sure daylight responsive controls are functioning.

6) IEQc8.1: Daylight and Views: Make sure project meets daylighting requirements.
7) IDc1, if applicable.

c. Mechanical and passive HVAC and refrigeration systems and related controls:

8) EAp2: Minimum Energy Performance: Confirm the project has met the mandatory provisions and prescriptive requirements of ASHRAE 90.1-2007.
9) EAp3: Fundamental Refrigeration Management: no CFCs in newly purchased equipment.
10) EAc1.3, Ch1: Optimize Energy Performance-HVAC: Confirm HVAC equipment has been sized per actual loads, meets enhanced efficiency standards, has proper controls and zoning, and uses variable speed controls.
11) EAc1.3, Ch2: Same as EAp2.

12) IEQp1: Minimum IAQ Performance: Confirm compliance of ANSI/ASHRAE 62.1-2007 for ventilation.
13) IEQp2, Ch2: Environmental Tobacco Smoke (ETS) Control: Verify requirements for designated smoking rooms
14) IEQp2, Ch3: Environmental Tobacco Smoke (ETS) Control: air sampling and air leakage test results for residential facilities.
15) IEQc1: Outdoor Air Delivery Monitoring: Confirm monitoring system is functioning.
16) IEQc2: Increased Ventilation: Passive systems: confirm building meets minimum flow rates. Mechanical systems: Confirm air testing, balance, and increase ventilation rates.
17) IEQc3.1: Construction IAQ Management Plan: During Construction: Confirm the replacement of proper filtration media (min. MERV 8).
18) IEQc5: Indoor Chemical and Pollutant Source Control Products: Confirm the exhaust system for spaces with hazardous gases or chemicals is working, the use of MERV 13 air filtration media.
19) IEQc6.2: Controllability of Systems: Thermal Comfort: Confirm the controls for ventilation and temperature are working properly.
20) IEQc7.1: Thermal Comfort: Design: Confirm control systems and HVAC system meet ANSI/ASHRAE 55-2004 requirements.

21) IEQc7.2: Thermal Comfort: Verification: Confirm monitoring system is working properly.
22) IDc1, if applicable.

d. Domestic hot water systems (Plumbing):

- EAp2: Minimum Energy Performance: Confirm the project has met the mandatory provisions and prescriptive requirements of ASHRAE 90.1-2007.
- IDc1, if applicable.

Mnemonics: REMP (See bold and underlined letters for items a, b, c, and d above)

Fundamental commissioning may also relate to the following credits:

- WEc1: Water Use Reduction
- EAc3: Measurement and Verification

EAp1 defines a minimum threshold for commissioning. The GBCI awards additional verification and rigor in the following related credit:

- EAc2: Enhanced Commissioning

Possible Strategies and Technologies:

Encourage owners to seek qualified individuals with high levels of experience in the following areas to lead the commissioning process:

a. Process management and commissioning planning.
b. The design, installation, and operation of the energy system.
c. Knowledge of energy systems' automation control.
d. Practical experience with start-up, performance, interaction, troubleshooting, balancing, testing, maintenance, and operating procedures of the energy systems.

Encourage owners to include building envelope systems, water-using systems, etc., in the commissioning plan if applicable. A proper building envelope can reduce energy consumption and improve indoor air quality and occupant comfort. Owners can save a lot of money and improve indoor air quality if they include building envelope commissioning, even though it is not required to be commissioned by LEED.

Refer to the USGBC reference guide for guidance on:

a. Commissioning reports
b. Commissioning plans
c. Commissioning specifications
d. Basis of Design (BOD)
e. Owner's Project Requirements (OPR)
f. Performance verification documentation

For LEED CI, the scope of commissioning can vary greatly: some may only include lighting system, while others may include all EMP (Electrical: Lighting, Mechanical: HVAC, and Plumbing: service water) systems.

Extra Credit (Exemplary Performance):
None

Project Phase:
Construction Administration (Commissioning involves many project phases, but a majority of its activities happens in Construction Administration phase).

LEED Submittal Phase:
Construction

Related Code or Standard:
None

Responsible Party:
Commissioning Authority/Agent (CxA), Owner, and Contractor

EAp2: Minimum Energy Performance (Mandatory, 0 points)

Purpose:

1) To set up the minimum energy efficiency level for the tenant space systems.
2) To avoid excessive energy use and related economic and environmental impact.

Credit Paths (0 points, mandatory requirements/prerequisites):

Your design for the tenant's work must meet ANSI/ASHRAE/IESNA Standard 90.1-2007 (with errata but without addenda), and the following:

a. Compared with the baseline per ANSI/ASHRAE/IESNA Standard 90.1-2007 (with errata but without addenda), you need to show a 10% reduction for the connected lighting power density for the tenant space. This can be achieved by applying the whole building lighting power allowance to the entire tenant space, or using the Space-by-Space method.
b. Meet the **mandatory provisions** (sections 5.4, 6.4, 7.4, 8.4, 9.4 and 10.4) in ASHRAE/IESNA Standard 90.1-2007 (with errata but without addenda).
c. Meet the **performance requirements** (section 11), or the **prescriptive requirements** (sections 5.5, 6.5, 7.5, and 9.5) of ANSI/ASHRAE/IESNA Standard 90.1-2007 (with errata but without addenda).
d. Based on rated-power, the project team shall install ENERGY STAR® qualified equipment for half of the ENERGY STAR® eligible equipment within the scope of work for the tenant. This applies to appliances, commercial food service equipment, electronics, and office equipment, but does NOT apply to lighting, HVAC, and building envelope products.

Two important definitions related to this prerequisite:
Process energy: Refrigeration and kitchen cooking, laundry (washing and drying), elevators and escalators, computers, office and general miscellaneous equipment, lighting not included in the lighting power allowance (such as lighting as part of the medical equipment), and other uses like water pumps, etc.

Regulated (non-process) energy: HVAC, exhaust fans and hood, lighting for interiors, surface parking, garage parking, building façade and grounds, space heating and service water heating, etc.

To meet EAp2, use the same process loads for your proposed building performance rating and the baseline performance building rating. You can use the Exceptional Calculation Method (ANSI/ASHRAE/IESNA 90.1-2007 G2.5) to record measures of reducing process loads. Include assumptions made for both your proposed design and the base building, and related supporting empirical and theoretical information in your documentation of process load energy savings.

If your project is in California, you can use Title 24-2005, Part 6 instead of ANSI/ASHRAE/IESNA 90.1-2007.

Considerations and Benefits:

1) **Environmental:** Energy used by the commercial real estate industry generates about 18% of the CO_2 emitted in the US. Buildings account for 48% of the energy used and related green house gas emissions.

2) **Financial:** Optimizing energy performance can reduce operating costs. This can be achieved with little or no cost. For example, changing an incandescent lamp to a fluorescent lamp can achieve a 75% reduction in energy use and save over $30 over the life span of the lamp.

Submittals:

1) Confirm the project meets ASHRAE (including addenda) requirements, and keep copies of related forms.
2) When using the **prescriptive compliance** path, gather paperwork showing that the project meets all requirements.
3) Confirm the project meets lighting power density reduction, and refer to EAc1.1 for more information on submittals.
4) Confirm the project meets ENERGY STAR® equipment requirements, and refer to EAc1.4 for more information on submittals.

Synergies:

Most LEED systems cover building energy efficiency in two places:

- EAp2: Minimum Energy Performance
- EAc1: Optimize Energy Performance

You can reduce energy consumption for a building by exceeding the building code requirements for HVAC systems, lighting, and the building envelope. You can also optimize the exterior lighting or use roofing materials suitable to the local climate to reduce energy use. Refer to the following credits (the bold credits are listed in USGBC Green Interior Design and Construction Reference Guide, and the remaining credits are listed in USGBC Green Building Design and Construction Reference Guide. I choose to list all of them because they are important and related to EAp2):

- **EAc1: Optimize Energy Performance**
- SSc1, Ch2F5: Heat Island Effect: Roof
- SSc1, Ch2F6: Light Pollution Reduction

Renewable energy may contribute to energy savings:

- SSc1, Ch2F11: On-Site Renewable Energy
- **EAc4: Green Power**

The project team must carefully coordinate IEQ and building energy performance. Do not compromise the well-being or health of the building occupants to save energy. You may need to use extra energy to increase ventilation, and the extra energy may cause water and air pollution. You may use economizer and/or heat-recovery ventilation to reduce energy use. Review and coordinate these credits carefully:

- IEQp1: Minimum IAQ Performance
- IEQc1: Outdoor Air Delivery Monitoring
- IEQc2: Increased Ventilation
- IEQc6: Controllability of Systems
- IEQc7: Thermal Comfort
- IEQc8: Daylight and Views

Reducing water use, especially hot water use, may save energy. Refer to the following credits for strategies:

- WEp1: Water Use Reduction
- WEc1: Water Use Reduction

Possible Strategies and Technologies:

1) The standards for LEED CI only apply to the tenant's scope of work. This will allow the tenant to seek LEED certification without forcing the landlord to make change to existing building system. However, the project still needs to meet ASHRAE 90.1-2007 (with errata but without addenda). Do not draft your lease to outline the tenant scope of work to avoid these standards.
2) Meet the local code if it is more restrictive than ASHRAE 90.1-2007 (with errata but without addenda), and has shown textural and quantitative equivalence per the U.S. Department of Energy (DOE) standard process for commercial energy code determination. Refer to the following link for more information on the DOE process for commercial energy code determination:
   http://www.energycodes.gov

   California Title 24 is accepted by USGBC as more restrictive than ASHRAE 90.1-2007 (with errata but without addenda).

Extra Credit (Exemplary Performance):

None

Project Phase:

Schematic Design

LEED Submittal Phase:

Design

Related Code or Standard:

1) ANSI/ASHRAE/IESNA Standard 90.1-2007: Energy Standard for Buildings Except Low-Rise Residential

2) Energy Star Program, Target Finder Rating Tool

Responsible Party:
MEP Engineer and Owner

EAp3: Fundamental Refrigerant Management (Mandatory, 0 points)

Purpose:
To minimize the depletion of stratospheric ozone.

Credit Paths (0 point, mandatory requirements/prerequisites):

1) The HVAC&R (R is for Refrigeration) system of your new base building should NOT use CFC-based refrigerants.

Note:

1) Do NOT include HVAC units with less than 0.5lb of refrigerant and small water coolers and standard refrigerators in the base building calculation.
2) Select the HVAC units with small ODP (Ozone Depletion Potential), GWP (Global Warming Potential), and short environmental life.
3) HVAC units with payback times longer than 10-years are non-feasible units.
4) The manufacture of HVAC units containing CFCs was stopped in 1995 in the US and these units will be phased out from existing buildings in the U.S. by 2011.

Considerations and Benefits:

1) **Environmental:** CFCs cause severe damage to the ozone layer. The ozone layer helps the stratosphere to absorb part of the sun's ultraviolet radiation, such that it does not reach the earth's surface. Avoiding the use of CFCs can help to protect the ozone layer.

2) **Financial:** All new equipment for HVAC&R systems shall have no CFC-based refrigerants. If the energy savings from replacing or upgrading old equipment with CFC-based refrigerants can offset the cost, then a CFC phase-out plan has to be implemented to earn this prerequisite. If the cost cannot be offset, then the project team must provide an audit by a qualified third party to verify that the CFC replacement or conversion is NOT economically feasible.

Submittals:

1) For major renovations, create and track the phase-out plan.
2) Gather manufacturer's data showing the type of refrigerant used by the base building HVAC&R system in the tenant's scope of work.

Synergies:
EAp3 has the minimum requirements for refrigerant selection. Refer to "EAc4: Enhanced Refrigerant Management" for further requirements.

Possible Strategies and Technologies:

1) Use no CFC-based refrigerants in new buildings.
2) For HVAC systems in existing buildings, create an inventory listing the equipment using CFC

refrigerants and supply a schedule to replace these refrigerants.

Extra Credit (Exemplary Performance):
None

Project Phase:
Design Development

LEED Submittal Phase:
Design

Related Code or Standard:
1) U.S. EPA Clean Air Act, Title VI, Section 608, Compliance with the Section 608 Refrigerant Recycling Rule
2) Montreal Protocol 1987

Responsible Party:
MEP Engineer and Owner

EAc1.1: Optimize Energy Performance: Lighting Power (1-5 points)

Purpose:
1) To avoid excessive energy use and reduce related economic and environmental impacts, and
2) To achieve improved energy performance above the baseline in the prerequisite standard.

Credit Path:
Compared with the baseline per ANSI/ASHRAE/IESNA Standard 90.1-2007 (with errata but without addenda), you need to show a certain percentage reduction for the connected lighting power density for the tenant space. This can be achieved by applying the whole building lighting power allowance to the entire tenant space, or by using the Space-by-Space method.

Points can be earned per the table below:

Percentage Reduction for the Connected Lighting Power Density	Points
15%	1
20%	2
25%	3
30%	4
35%	5

If your project is in California, you can use Title 24-2005, Part 6 instead of ANSI/ASHRAE/IESNA 90.1-2007.

Considerations and Benefits:
1) **Environmental:** Energy used by the commercial real estate industry generates about 18% of the

CO_2 emitted in the US. Buildings account for 48% of the energy used and the related greenhouse gas emission.

The traditional way of producing energy has a negative environmental impact: fossil fuels generate air and water pollution. Other sources of energy can have environmental impacts as well. For example, hydropower can change aquatic ecosystems and have negative impacts on endangered species. Nuclear power plants can generate radioactive waste. Thus, saving energy can mitigate negative environmental impacts.

2) **Financial:** Optimizing energy performance can reduce operating costs, resulting in smaller equipment and reduced required space for equipment. This can be achieved with little or no cost. For example, changing an incandescent lamp to a fluorescent lamp can achieve a 75% reduction in energy use and save over $30 over the life span of the lamp.

Submittals:

1) Confirm the project meets ASHRAE (including addenda) requirements, list addenda used, and keep copies of related forms.
2) Document the rated power of lighting systems installed in the project, and confirm the project's lighting power density reduction.
3) Collect compliance paperwork for ANSI/ASHRAE/IESNA 90.1-2007 user's manual.
4) Document types of space area or building area with the related gross floor areas.
5) Collect information on extra lighting power allowances, and confirm they only apply to lighting systems separated from general lighting.

Synergies:

Most LEED systems cover building energy efficiency in two places:

- EAp2: Minimum Energy Performance
- EAc1: Optimize Energy Performance

You can reduce lighting power consumption for a building by using daylighting or task lights, selecting proper target light levels, and letting occupants control their lighting levels to increase their satisfaction. Refer to the following prerequisites and credits (the bold credits are listed in USGBC Green Interior Design and Construction Reference Guide, and the remaining credits not listed in the reference guide, but are related to EAc1.1. I choose to list all of them because they are important):

- **EAp2: Minimum Energy Performance**
- **EAc1.2: Optimize Energy Performance: Lighting Controls**
- SSc1, Ch2F6: Light Pollution Reduction

Renewable energy may contribute to energy savings:

- SSc1, Ch2F11: On-Site Renewable Energy
- EAc4: Green Power

The project team must carefully coordinate IEQ and building energy performance. Do not compromise the well-being or health of the building occupants to save energy. Review and coordinate these credits carefully:

- **IEQc6.1: Controllability of Systems: Lighting**
- **IEQc8.1: Daylight and Views: Daylight**

The project team must use commissioning to make sure light controls work properly. Refer to the following prerequisites and credits:

- **EAp1: Fundamental Commissioning of the Building Energy Systems**
- **EAc2: Enhanced Commissioning**

Possible Strategies and Technologies:

1) With the help of a computer simulation model, you can quantify energy performance by comparing the model to the baseline building, assessing your building's energy performance, finding out the most cost-effective energy efficiency measures, and designing your building systems and envelope to achieve the best energy performance.
2) The project team shall balance the various needs of the space when using task lighting.
3) Lighting can have on, off, and dimming modes to fully take advantage of daylighting.
4) Apply highly reflective internal surfaces and high-efficacy sources to reduce the lighting power density.

Extra Credit (Exemplary Performance):

The project team can earn one extra point by reducing the lighting density to 40% below the standard.

Project Phase:

Schematic Design

LEED Submittal Phase:

Design

Related Code or Standard:

1) ANSI/ASHRAE/IESNA Standard 90.1-2007: Energy Standard for Buildings Except Low-Rise Residential
2) U.S. Department of Energy

Responsible Party:

MEP Engineer and Owner

EAc1.2: Optimize Energy Performance: Lighting Controls (1-3 points)

Purpose:

1) To avoid excessive energy use and reduce related economic and environmental impacts, and
2) To achieve improved energy performance beyond the baseline in the prerequisite standard.

Credit Path:

Design your project to include at least one of the following features:

1) **In daylit areas, add daylight controls: (1 point)**
   Within 15 feet of windows and under skylights, add daylight controls, which must dim or switch electric lights based on the presence or absence of daylight.
2) **Add daylight controls for at least 50% of the lighting load: (1 point)**
   Add and demonstrate daylight responsive controls for at least 50% of the connected lighting load. The controls must dim or switch electric lights based on the presence or absence of daylight.

3) **Add occupancy sensors: (1 point)**
   Add occupancy sensors for at least 75% of the connected lighting load.

Considerations and Benefits:

Most occupants like daylighting, which reduces or eliminates electric lighting, saves energy, and provides an improved indoor environment. Daylight responsive controls can maximize the benefits of daylighting. When daylighting is coordinated with HVAC systems, the net cost of utilities can be reduced.

Refer to EAc1.1 for more information on Considerations and Benefits.

Submittals:

1) Highlight daylit zone on the lighting plan; show which lights are linked to each control.
2) For more complex lighting plan, add a lighting control summary showing the combined control logic.

Synergies:

Refer to EAc1.1

Possible Strategies and Technologies:

1) Daylight responsive controls shall be a part of an overall design strategy: utilizing the least possible electric lighting, optimizing natural daylight, incorporating proper task lighting, and including controllability of lighting by building occupants.
2) Use both daylight responsive controls and occupant sensors. Daylight responsive controls normally have a photosensor to adjust the electric power required.
3) The project team shall balance the various needs of the space when using task lighting.
4) Lighting can have on, off, and dimming modes to fully take advantage of daylighting. Continuous dimming is the best choice, followed by step-dimming, bi-level, and on-off controls.
5) Apply highly reflective internal surfaces and high-efficacy sources to reduce the lighting power density.

Extra Credit (Exemplary Performance):

The project team can earn one extra point by using daylight-responsive controls for 75% of the connected lighting load, OR by using occupancy-responsive controls for 95% of the connected lighting load.

Project Phase:

Schematic Design

LEED Submittal Phase:

Design

Related Code or Standard:

None

Responsible Party:

MEP Engineer and Owner

EAc1.3: Optimize Energy Performance: HVAC (5-10 points)

Purpose:

1) To avoid excessive energy use and reduce related economic and environmental impacts, and
2) To achieve improved energy performance above the baseline in the prerequisite standard.

Credit Path:

Choice 1: Select at least one of the following strategies:

1) **Equipment Efficiency (5 points)**
   Install HVAC systems that meet the criteria of New Building Institute's **Advanced Buildings™ Core Performance™ Guide**, Section 1.4: Mechanical System Design, 2.9: Mechanical Equipment Efficiency, and 3.10: Variable Speed Control.

2) **Proper Zoning and Controls (5 points)**
   Zone tenant space to meet the following:
   - Each solar exposure shall have a separate control zone.
   - Interior spaces shall be separately zoned.
   - Special occupancies (kitchen, conference rooms, etc.) and private offices shall have controls with sensors. Adjust the HVAC systems per space demand and use.

OR

Choice 2: Reduce design energy cost when compared with the cost budget for regulated energy components per the baseline of ANSI/ASHRAE/IESNA Standard 90.1-2007 (with errata but without addenda)

AND

1st Path (5 points)

Show the criteria for tenant space HVAC system are 15% better than a system per the baseline of ANSI/ASHRAE/IESNA Standard 90.1-2007 (with errata but without addenda)

OR

2nd Path (10 points)

Show the criteria for tenant space HVAC system are 30% better than a system per the baseline of ANSI/ASHRAE/IESNA Standard 90.1-2007 (with errata but without addenda)

Note:

1) The project team shall only consider the energy costs for space cooling, space heating, and related pumps and fans.
2) If the project team wants to use the addenda approved by ASHRAE, then they must be used consistently for all LEED credits.

Considerations and Benefits:

Refer to EAc1.1

Submittals:

1) A list of the functions and locations of HVAC system sensors or automatic controls

2) Paperwork showing the motives for zone distribution and possible energy savings for each control
3) To comply with the LEED cost-budget requirements, document the end use, cost, and related peak demand for the proposed design energy, as well as the baseline energy type and related cost.

Synergies:

EAc1.3 is related to the following prerequisite and credits:
IEQp1: Minimum IAQ Performance
IEQc1: Outdoor Air Delivery Monitoring
IEQc2: Increased Ventilation
IEQc6.2: Controllability of Systems: Thermal Comfort
IEQc7.1: Thermal Comfort: Design
IEQc7.2: Thermal Comfort: Verification

Possible Strategies and Technologies:

1) Plants provide shade, reduce heat island effect, protect buildings from cold north winds in winter, and save energy.
2) Orienting the long side of a building along an east-west axis can provide the most effective exterior shade, taking advantage of the dominant south wind in summer, and saving energy.
3) Confirm the building meets minimum insulation criteria per codes, and is weather-tight.
4) Confirm the landlord has set lighting density criteria for all tenant space. Lighting consumes a large portion of the building's energy and reducing common area lighting cost is an efficient way to save energy.
5) Take full advantage of natural light and natural ventilation: use light shelves, clerestories, high windows, fiber optic devices, and light pipes.
6) Inspect existing equipment and use high-efficiency HVAC equipment, fans, boilers, chillers, and pumps whenever possible.
7) Confirm the building has a functional building energy management system.

Extra Credit (Exemplary Performance):

With regard to the 2nd choice, the project team can earn one extra point by showing that the criteria for the tenant space HVAC system are 33% better than a system per the baseline of ANSI/ASHRAE/IESNA Standard 90.1-2007 (with errata but without addenda).

Project Phase:

Schematic Design

LEED Submittal Phase:

Design

Related Code or Standard:

1) ANSI/ASHRAE/IESNA Standard 90.1-2007 (with errata but without addenda): Energy Standard for Buildings Except Low-Rise Residential
2) New Building Institute's Advanced Buildings™ Core Performance™ Guide
3) U.S. Department of Energy

Responsible Party:

MEP Engineer and Owner

EAc1.4: Optimize Energy Performance: Equipment and Appliances (1-4 points)

Purpose:

1) To avoid excessive energy use, as well as reduce related economic and environmental impacts, and
2) To achieve improved energy performance above the baseline in the prerequisite standard.

Credit Path:

You can gain points per the following table:

% installed Energy Star Qualified Equipment of Energy Star Eligible Equipment	Points
70%	1
77%	2
84%	3
90%	4

Note: The project team shall include all appliances, including commercial food service equipment, appliance, electronics, and office equipment, excluding building envelope products, lighting, and HVAC.

Considerations and Benefits:

1) **Environmental:** Plug loads consume 9% of office building energy and 15% of electricity. These loads are hard to quantify and regulate. Appendix G of ANSI/ASHRAE/IESNA Standard 90.1-2007 (with errata but without addenda) uses a performance rating to address plug loads.

2) **Financial:** Energy Star qualified equipment reduced electricity use by 30% to 75% when compared with other equipment. The initial cost may be higher, but long term savings are substantial: If all businesses in the United States buy only Energy Star qualified computers this year, the savings is $260 million over the lifetime of these machines. That is enough to provide lighting for 130 million square feet of commercial buildings in the U.S. for a year.

Submittals:

1) A list of generic and Energy Star qualified equipment, related rated power (watts) and daily energy use for the tenant space

Synergies:

EAc1.4 is related to the following prerequisite and credits:
EAp1: Fundamental Commissioning of the Building Energy Systems
EAp2: Minimum Energy Performance
EAc1.3: Optimize Energy Performance: HVAC
EAc2: Enhanced Commissioning
IEQc7.1: Thermal Comfort: Design

Possible Strategies and Technologies:

1) Variations in work schedules and occupant densities make it difficult to set a constant metric for plug loads. For example, fast food restaurants rate poorly in plug load calculation even they may be very energy efficient. BTU per meal may be a better metric for them. Similarly, for retail space, number of customers served or sales volume correlates better to process load.
2) Using Energy Star qualified equipment is the most straightforward way to optimize energy performance.

Extra Credit (Exemplary Performance):

The project team can earn one extra point by achieving a rated power of 97% from Energy Star qualified equipment.

Project Phase:

Operation and Maintenance

LEED Submittal Phase:

Design

Related Code or Standard:

1) Energy Star Qualified Products
   http://www.energystar.gov
2) U.S. Department of Energy

Responsible Party:

MEP Engineer and Owner

EAc2: Enhanced Commissioning (5 points)

Purpose:

To designate an independent Commissioning Authority (CxA) before the start of the construction documents (CD) phase, and verify the tenant space systems' performance, and then perform additional activities.

Credit Path:

To achieve this credit, the project team needs to:

a. Perform ALL requirements of EAp1

AND

b. Perform (or have a contract to perform) the following commissioning process activities that are above and beyond the requirements of EAp1 and comply with the latest version of the related USGBC reference guide:

1) Perform a **commissioning documentation review:** Review contractor submittals related to systems commissioned to make sure they meet the **Basis of Design (BOD)** and **Owner's Project Requirements (OPR)**. This needs to be concurrent with Architect/Engineer reviews. The submittals shall also be sent to the owner and the design team.

2) Make a system manual to provide information for future system operators to be able to understand and optimally operate the commissioned systems.
3) Arrange to review building operations with occupants and O&M staff 8 to 10 months of the building's substantial completion. Make a plan to resolve outstanding commissioning-related issues.
4) At least one **design review** of **BOD** and **OPR** and design documents before the mid-construction documents phase, and back check subsequent design submissions against the review comments.
5) Confirm the completion of the requirements for training building occupants and operating personnel.

Note: The CxA shall be independent of the project's design and construction management teams, and he cannot be an employee of the design firm providing the services, though he can be contracted through the design firm. He cannot be an employee of, or contracted through the construction manager or contractor holding the construction contract (This is different than EAp1). The CxA could be a qualified consultant or employee of the owner.

Mnemonics: SMART (See underlined letters in items #1 through #5 above). The CxA shall perform the #1, #3 and #4 (or S, A and R) of above tasks; the other team member can perform #2 and #5 (or M and T) of the above tasks.

Considerations and Benefits:

1) **Environmental:** If building energy systems do not perform as designed, they may consume substantially more energy in a long term. Enhanced Commissioning can confirm a system's performance per the owner's requirements, reduce contractor callbacks, lower operating costs, reduce energy uses, result in better building documentation, and increase occupant productivity.

2) **Financial:** If the project team has not included commissioning as part of the budget from the very beginning, owners/ tenants may oppose it because of the cost. However, if you include long-term benefits, commissioning is very cost-effective. It can reduce energy cost, increase occupant productivity, reduce tenant vacancy and turn-over, and liabilities related to indoor air quality (IAQ), and prevent premature replacement of equipment.

Submittals:

All submittals required by EAp1 plus the following:

1) A written schedule of building operator training
2) A copy of the CxA's design review, any related designer response, and back-check confirmation

Synergies:

Same as EAp1, except that EAc2 goes above and beyond the minimum threshold for commissioning defined by EAp1.

Possible Strategies and Technologies:

The CxA can be contracted by

1) The owner (the preferred way)

OR

2) Through construction management firms NOT holding the construction contracts or through the design firm.

You can find detailed guidance on the rigor expected for the following items in the related reference guide:

a. Systems manual;
b. Commissioning design review;
c. Commissioning submittal review.

For EAc2, the CxA shall be involved with the project early and involved with a design review and a commissioning documentation review.

Extra Credit (Exemplary Performance):
None

Project Phase:
Construction Administration (Commissioning involves many project phases, but a majority of its activities happen in Construction Administration phase).

LEED Submittal Phase:
Construction

Related Code or Standard:
None

Responsible Party:
Commissioning Authority, Owner, and Contractor

EAc3: Measurement and Verification (2-5 points)

Purpose:
To have ongoing optimization and accountability of tenant water and energy consumption over time.

Credit Paths:

1st Case: Tenant Space Occupying Less than 75% of the Total Building Area:
Achieve at least one of the following:

- Use a sub-meter to measure and document energy use within the tenant space. (2 points)
- Negotiate a lease whereby the tenant's energy costs are separate from the base rent. (3 points)

OR

2nd Case: Tenant Space Occupying 75% or More of the Total Building Area:
Install continuous metering devices for the following: (5 points)

Mechanical (HVAC):

- Air and water economizer and heat recovery cycles
- Air distribution static pressure and ventilation air volumes
- Cooling load

Electrical:
- Lighting systems and controls

Plumbing:
- Indoor water riser and outdoor irrigation system
- Boiler efficiencies

Refrigeration:
- Chiller efficiency at variable loads (kW/ton)

Other:
- Constant and variable motor loads
- Variable frequency drive operation
- Building-related process energy and equipment

For 2nd case above, the project team needs to create and implement a Measurement and Verification (M&V) Plan that includes information above and is consistent with Option B, C, or D of the *International Performance Measure and Verification Protocol (IPMVP) Volume I: Concepts and Options for Determining Energy and Water Savings in New Construction*, 2001. (The 2009 edition is already available. I expect USGBC will update to the latest the edition very soon)

Offer a process for corrective action if the M&V plan results show the energy savings have not been achieved.

Submittals:
1) Create an IPMVP-compliant M&V plan, and confirm it is consistent with Option B, C, or D
2) For Tenant Space Occupying 75% or more of the total building area, collect utilities bills, lease agreements or other paperwork to show that tenant energy costs are paid by the tenant
3) Summarize the monitoring systems for the tenant space, and show that at least the required systems are monitored

Synergies:

Measurement and verification may help to achieve optimal energy performance and ensure accountability. If you need to obtain an energy performance contract or other funding, use the International Performance Measurement and Verification Protocol (IPMVP). M&V plans typically include tracking the performance of renewable energy generation systems to identify operational issues.

EAc3 may contribute to the following credits:
- SSc1, Ch2F11: On-Site Renewable Energy
- EAp2: Minimum Energy Performance
- EAc1: Optimize Energy Performance
- EAc4: Green Power

Commissioning and M&V often use the same devices. Refer to the following credits for related criteria:
- EAp1: Fundamental Commissioning of the Building Energy Systems
- EAc2: Enhanced Commissioning

Possible Strategies and Technologies:

1) Create an M&V Plan to evaluate energy systems and/or building performance.
2) Use engineering analysis and energy simulation to characterize the energy systems and/or building.
3) Use proper metering to measure the use of energy.
4) Verify performance by comparing actual performance to predicted performance **(Performance Factor)**; then group them by system or component if needed.
5) Compare actual performance to baseline performance to track energy **efficiency**.

IPMVP can be used to verify savings related to **energy conservation measures (ECMs)** and strategies. This LEED credit goes above and beyond normal IPMVP M&V goals. M&V can go above and beyond energy systems and ECMs, and energy conservation strategies. IPMVP gives guidance on M&V strategies and their uses. Use these together with trend logging and monitoring of important energy systems, and seek ongoing accountability for building energy performance.

Extra Credit (Exemplary Performance):
None

Project Phase:
Design Development

LEED Submittal Phase:
Design

Related Code or Standard:

1) International Performance Measure and Verification Protocol (IPMVP), Volume I, Efficiency Valuation Organization (EVO), Concepts and Options for Determining Energy Savings in New Construction, effective 2001

Responsible Party:
MEP Engineer, Building Controls Designer/Manufacturer

EAc4: Green Power (5 points)

Purpose:
To encourage the use of renewable, grid-source energy technologies based on zero net pollution.

Credit Paths:

1st Choice:
Obtain a minimum of 50% of your building's electricity (based on quantity, NOT the cost) from renewable sources through two-year or longer contracts of renewable energy. Use the definition of a renewable source by the Center for Resource Solution's (CRS) Green-e product requirements.

OR

2nd Choice:
Purchase a minimum of 8 kwh (kilowatt hours) per s.f. per year of your building's electricity (based on quantity, NOT the cost) from renewable sources through two-year or longer contracts of renewable energy. Use the definition of a renewable source by the Center for Resource Solution's (CRS) Green-e product requirements.

Considerations and Benefits:

1) **Environmental:** The traditional way of producing energy has negative environmental impacts: fossil fuels generate air and water pollution. Other sources of energy can have environmental impacts as well. For example, hydropower can change aquatic ecosystems and have negative impacts on endangered species. Nuclear power plants can generate radioactive waste.

   Green power products rely on renewable sources, such as biomass, geothermal wind, and solar sources. They can greatly alleviate the negative impact of the traditional energy. The Center for Resource Solution's (CRS) Green-e program establishes a nationally recognized and rigorous standard, and can help consumers identify green power products.

2) **Financial:** Green power products may have a higher initial cost, but they rely on stable and renewable sources. As the green power market matures, these costs are expected to fall. Various governing agencies may have incentives for using green power. Refer to the Database for State Incentives for Renewables and Efficiency: (DSIRE)
   http://www.dsireusa.org

Submittals:

1) Keep a record of the signed 2-year contract to purchase Green-e certified renewable energy or equivalent
2) Keep paperwork showing renewable energy was retained on behalf of the project when others are purchasing the renewable energy

Synergies:

Using renewable energy may reduce energy costs, refer to:

- EAc1: Optimize Energy Performance

Commission the on-site renewable energy system. Check the roof for structural stability before installing rooftop equipment.

EAc4 may contribute to the following prerequisite and credits:

- SSc1, Ch2F5: Heat Island Effect: Roof
- EAp1: Fundamental Commissioning of the Building Energy Systems
- EAc2: Enhanced Commissioning

Possible Strategies and Technologies:

1) Evaluate your building's energy needs and find opportunities to sign a green power contract.
2) Green power comes from wind, solar, biomass, geothermal, or low-impact hydro sources. See www.green-e.org for more information on the Green-e program.
3) The power you buy for this credit does NOT need to be Green-e certified. You can use green power from other sources as long as they meet the technical requirements of the Green-e program.
4) Use green tags, tradable renewable certificates (TRCs), renewable energy certificates (REGs),

and other green power sources to meet EAc4 requirements as long as they meet the technical requirements of the Green-e Program.

Extra Credit (Exemplary Performance):

You can get one innovation point by purchasing 100% (or 16 kwh per s.f. per year) of the building's electricity from renewable sources.

Project Phase:

Occupation/Operation

LEED Submittal Phase:

Construction

Related Code or Standard:

1) Center for Resource Solution (CRS), Green-e Product Certification Requirements
2) Department of Energy (DOE)
3) Commercial Buildings Energy Consumption Survey (CBECS).

Responsible Party:

LEED AP and Owner

EA Summary and Mnemonics:

EA Credit Name	Extra Credit	Responsible Party
EAp1: **F**undamental Commissioning of the Building Energy Systems	0	Commissioning Authority/Agent (CxA), Owner and Contractor
EAp2: **M**inimum Energy Performance	0	MEP Engineer and Owner
EAp3: **F**undamental Refrigerant Management	0	MEP Engineer and Owner
EAc1.1: **O**ptimize Energy Performance: Lighting Power (1-5 points)	The project team can earn one extra point by reducing the lighting density to 40% below the standard.	MEP Engineer and Owner
EAc1.2: Optimize Energy Performance: Lighting Controls (1-3 points)	The project team can earn one extra point by using daylight-responsive controls for 75% of the connected lighting load, OR by using occupancy-responsive controls for 95% of the connected lighting load.	MEP Engineer and Owner
EAc1.3: Optimize Energy Performance: HVAC (5-10 points)	For 2nd choice, the project team can earn one extra point by showing the criteria for tenant space HVAC system are 33% better than a system per the baseline of ANSI/ASHRAE/IESNA Standard 90.1-2007 (with errata but without addenda).	MEP Engineer and Owner
EAc1.4: Optimize Energy Performance: Equipment and Appliances (1-4 points)	The project team can earn one extra point by achieving a rated power of 97% from Energy Star qualified equipment.	Commissioning Authority, Owner, and Contractor
EAc2: **E**nhanced Commissioning	0	MEP Engineer, Building Controls Designer/Manufacturer
EAc3: **M**easurement and Verification (2-5 points)	0	MEP Engineer, Building Controls Designer/Manufacturer
EAc4: **G**reen Power (5 points)	You can get one innovation point by purchasing 100% (or 16 kwh per s.f. per year) of the building's electricity from renewable sources.	LEED AP and Owner

Mnemonics:

Find My Fun!

Oprah's Energy Management Group. (See bold and underlined letters in credit names on table above also). Of course, you can create your own **mnemonics** to help you memorize them.

D. Materials and Resources (MR)

Overall purpose: The MR category has many purposes including:

1) Use fewer materials (**Reduce**)
2) Use materials with less environmental impact
3) Manage and reduce waste (**Recycle**)
4) Storage and Collection of Recyclables
5) Building **Reuse**: Maintain Existing Walls, Floors and Roof
6) Building Reuse: Maintain Interior Nonstructural Elements
7) Construction Waste Management
8) Materials Reuse
9) Recycled Content
10) Regional Materials
11) Rapidly Renewable Materials
12) Certified Wood

Mnemonics:

My effortless work at SBC
MR Regan Ray Carter (See underlined letters above also)

Core concepts:

1) Manage waste
   - Reduce waste
   - Reuse and divert waste
2) Reduce and reuse materials
   - Reduce materials used
   - Reuse of building and materials
   - Choose rapidly renewable materials

Recognition, regulation and incentives:

1) Recognition
   - Cradle to Cradle, Green Seal, and other product certifications
2) Regulation (requirements and goals)
   - Rare
   - Internal policy for supply chain and materials management in some organizations
3) Financial incentives
   - Recycling incentive

Overall strategies and technologies:

Note: Not **all** strategies and technologies have to be used simultaneously for your project

1) **Reduce** waste
2) Purchase sustainable materials
3) **Reuse** and divert waste: **recycle** solid waste and demolition waste
4) Reduce life cycle impact
5) Reduce demand for materials: Implement in design and construction; use new technologies
6) Reuse all or a portion of the existing building
7) Reuse materials: Use refurbished, salvaged, and reclaimed materials; purchase refurbished and

reclaimed materials

8) Use **rapidly renewable materials**: wool carpeting, cork flooring, sunflower seed board panels, linoleum flooring, bamboo flooring, cotton batt insulation
Mnemonics:
WC on slab (See underlined letters above also)
9) Choose materials with a reduced life cycle impact: regional materials; certified wood; materials containing pre- and post-consumer recycled content

Specific Technical Information:

MRp1: Storage and Collection of Recyclables (Mandatory, 0 points)

Purpose:
To reduce waste generated by building occupants that is going to landfills.

Credit Path (0 point, mandatory requirements/prerequisites):
Provide an easily accessible designated area for separation, collection, storage, and recycling of non-hazardous materials, like **p**aper, corrugated **c**ardboard, **m**etals, **g**lass, and **p**lastics for the tenant space.

Mnemonics: People Can Make Green Promises (See **bold** and underlined letters at the last part of the sentence listing the recycled materials. I use bold font here also because the underline for letter "g" is hard to read.)

Note: These five materials required for recycling (as listed above) make up 59% of the total municipal solid waste stream.

Food scraps (12%) and yard trimmings (13%) make up for 25% of the total. The GBCI encourages you to compost these types of waste on-site if possible.

The remaining waste is wood (6%), textiles, leather and rubber (7%), and other (3%).

Considerations and Benefits:

1) **Environmental:** An easily accessible designated area for separation, collection, storage, and recycling can help to divert a large portion of solid waste stream from the landfill. Recycling saves virgin natural resources, as well as reduces air, water, and land pollution. An educational program for building occupants can encourage them to participate.

2) **Financial:** Recycling facilities, such as recycling bins and storage, can increase project's initial costs and take up area, but it can also save disposal and tipping fees. For large projects, cardboard balers, can crushers, or other equipment can reduce the recycling area. Recycling can save money for a project over the long term.

Community-based recycling programs can increase recycling rates, and create a more stable recycling market.

Submittals:

1) Attain a letter from the landlord outlining a recycling program/plan showing the recycling area's size and accessibility to facility, staff, and occupants. Provide adequate size.
2) Create floor plans, site plans, etc. to highlight all storage areas.

Synergies:

1) The project team can use displays and signage to inform visitors and occupants about on-site recycling and seek an ID credit for educational outreach.
2) Select a LEED-Certified building because it probably already has a recycling program.
3) Follow tenant guidelines to address recycling policy and procedure. Encourage the **3 R's**: reduce, reuse, and recycle.

Possible Strategies and Technologies:

1) When selecting a tenant space, coordinate with all collection services to assure effective size and function of the designated recycle areas.
2) Use these methods: separation, collection, and storage.
3) Use aluminum can crushers, cardboard balers, collection bins and recycling chutes (collection sources) to make the recycling program more effective.
4) Select a LEED-Certified building, follow the building's recycling policy and tenant guidelines, and encourage occupants to reduce and reuse before recycling. You also can consider the waste management and maintenance practices for the entire building.

   If a landlord is providing cleaning services for all tenants, then he can control the procedures and space needed for storing, removing, and hauling recycled materials. He needs to coordinate the space needed and the frequency of collecting the recycled materials and waste from the tenant space.

   If tenants are handling their own cleaning services, then the landlord needs to provide adequate space for storing recycled materials and clear instructions in the tenant guidelines regarding use of the space. The landlord may provide several collection points at the central building area for tenants' convenience in large buildings or buildings with multiple floors.

5) **Guideline for Recycling Area**

Commercial Building Area (s.f.)	Min. Recycling Area (s.f.)
0-5,000	82
5,001-15,000	125
15,001-50,000	175
50,001-100,000	225
100,001-200,000	275
200,000 or more	500

Extra Credit (Exemplary Performance):

None

Project Phase:
Schematic Design

LEED Submittal Phase:
Design

Related Code or Standard:
California Integrated Waste Management Board 1999 (CIWMB 1999)

Responsible Party:
Architect and Owner

MRc1.1: Tenant Space: Long-Term Commitment (1 point)

Purpose:
To reduce tenant's waste and the related negative environmental impacts in transportation, manufacturing, and materials, as well as to promote resource conservation.

Credit Path:
The tenant must stay in the same space for at least 10 years. (1 point)

Considerations and Benefits:

1) **Environmental:** Long-term leases reduce the frequency of moving and related tenant improvement work and increase the return on energy efficiency improvements. Further, occupant quality of life, employee convenience and commute time, and related issues affect the choice of tenant space. Cultural amenities, mass transit, and other community improvements take time to develop. Long-term leases benefit the entire community.

2) **Financial:** Moving involves substantial costs, such as moving equipment and labor, employee downtime, etc. Long-term leases guarantee the space for the tenant for a long time, and reduce moving costs.

Submittals:
Keep a copy of the lease showing at least a 10-year term.

Synergies:
MRc1.1 is related to the following credits:

- SSc1: Site Selection
- SSc3.1: Alternative Transportation: Public Transportation Access
- EAc1: Optimize Energy Performance
- MRc3: Materials Reuse
- MRc4: Recycled Content
- MRc5: Regional Materials

- MRc6: Rapidly Renewable Materials
- MRc7: Certified Wood

- IEQc1: Outdoor Air Delivery Monitoring
- IEQc2: Increased Ventilation
- IEQc3: Construction IAQ Management Plan
- IEQc4: Low-Emitting Materials
- IEQc5: Indoor Chemical and Pollutant Source Control Products
- IEQc6: Controllability of Systems
- IEQc7: Thermal Comfort
- IEQc8: Daylight and Views

Possible Strategies and Technologies:

1) Sign a lease with at least a 10-year term.
2) Own the space: condominium ownership can satisfy this credit.
3) Select a site with potential space for growth.

Extra Credit (Exemplary Performance):

None

Project Phase:

Pre-Design

LEED Submittal Phase:

Design

Related Code or Standard:

None

Responsible Party:

Architect and Owner

MRc1.2: Building Reuse: Maintain Interior Nonstructural Components (1-2 points)

Purpose:

To fully take advantage of existing buildings, save resources, preserve cultural heritage, minimize waste and new buildings' environmental impact, and minimize material transportation and manufacturing costs.

Credit Path:

Keep 40% (1 point) or 60% (2 points) of existing nonstructural elements such as doors, floor coverings, ceiling systems, and interior walls (based on surface area).

If you are constructing an addition and the area of the new addition is more than two times the area of the existing building, your project is not eligible for this credit.

Do not include MEP units and elevator machines in the calculation.

Considerations and Benefits:

1) **Environmental:** Building reuse saves costs involving demolition, extracting raw materials, transportation, and manufacturing. It can reduce construction waste, preserve open space, historic buildings, and neighborhoods as well as reduce waste and emissions.

2) **Financial:** Building reuse may increase the soft cost of design because of its overall complexity, but it also can save substantial amounts of construction hard cost. Overall, the owner will likely save money.

Submittals:

1) A list of interior nonstructural elements, including IDs for elements and the total areas for reused, existing, and new elements

Synergies:

MRc1.2 is related to the following credits:

- MRc2: Construction Waste Management
- MRc3: Materials Reuse

If there are not enough reuse materials to achieve MRc1.2, then you can count the materials toward MRc2 or MRc3 (but not both).

Possible Strategies and Technologies:

1) Keep existing buildings when possible, but remove hazardous materials.
2) Document existing elements to be reused, detailing enough information to determine the surface area.
3) Division 12 items, furniture, and furnishing, are covered in MRc3.2: Materials Reuse: 30% Furniture and Furnishing.
4) If demolition is included in the building owner's scope of work, it defeats the purpose of this credit.

Extra Credit (Exemplary Performance):

The project team can obtain one extra ID point for keeping 80% of the existing nonstructural elements like doors, floor coverings, ceiling systems, and interior walls (based on surface area).

Project Phase:

Pre-Design

LEED Submittal Phase:

Construction

Related Code or Standard:

None

Responsible Party:

Architect, Contractor, and Owner

MRc2: Construction Waste Management (1-2 points)

Purpose:

1) To divert land-clearing, demolition, and construction debris from disposal in incinerators and landfills.
2) To redirect reusable materials to proper sites, and recover recyclable resources to be reused in the manufacturing process.

Credit Path for NC, CS, and Schools:

Salvage and/or recycle non-hazardous demolition and construction debris. Create and implement a plan to identify the materials to be reused, recycled, and recovered from disposal and whether they will be comingled or sorted on-site. You cannot count land-clearing debris and excavated soils as part of the material for this credit. You can calculate by volume or weight, but you need to be consistent throughout the whole job.

% of Salvaged or Recycled Materials by Weight or Volume	Points
50%	1
75%	2

Considerations and Benefits:

1) **Environmental:** Demolition and construction produce enormous amounts of waste: 136 million tons in 1996 alone in the US, 57% from nonresidential projects. 2 to 2.5 pounds of solid waste will be generated for each s.f. of commercial space, most of it can be recycled.

   Construction Waste Management reduces the needs for landfills, saves disposal fees, and costs for extracting raw materials, transportation, and manufacturing. It can divert construction waste, preserve open space, and reduce waste and emissions.

2) **Financial:** Construction Waste Management may increase the complexity of the project and initial cost, but it can also save a substantial amount of construction hard cost in. Overall, the owner will probably save money.

   Recyclable materials have different values. Overall, it is cost-effective to recycle cardboard, metals, concrete, and asphalt. Saving disposal fees is also an important incentive.

Submittals:

1) A log of all types of construction waste, the quantities of each type that were landfilled or diverted, and percentage of waste diverted
2) A construction waste management plan, including diversion goals, related construction materials and debris to be diverted, protocols for implementation, and responsible parties

Synergies:

MRc2 is related to the following credits:

- MRc1.2: Building Reuse: Maintain Interior Nonstructural Components
- MRc3: Materials Reuse

If there are not enough reuse materials to achieve MRc1.2, then you can count the material toward MRc2 or MRc3 (but not both).

If your project has asbestos, lead, or other contaminated substances, they should be remediated per EPA requirements. Refer to the following credit:

- SSc1, Ch2F1: Brownfield Redevelopment

Possible Strategies and Technologies:

1) Set goals for diverting waste and create and implement a construction management plan to achieve these goals.
2) Many building materials can be salvaged, refurbished, reused, or recycled: glass, plastic, wood, metal, brick, cardboard, insulation, carpet, acoustic ceiling tiles, and gypsum wallboard, etc.
3) Salvageable materials include: beams, cabinetry, flooring, and paneling.
4) Include on-site salvaged and reused materials if they have NOT been used for MRc3.
5) Select an area of your job site for comingled or segregated materials for recycling; keep track of your diverting and recycling effort.
6) Select recyclers and haulers to handle recycled materials.
7) Salvage your materials on-site or donate your materials to charity.
8) Exclude soils, vegetation, and rocks from the site, but include MEP equipment (Note: MEP equipment can ONLY be included in MRc2 or MR c7).

Extra Credit (Exemplary Performance):

You can get one innovation point for MRc2 by salvaging and/or recycling a minimum of 95% of total construction debris.

Project Phase:

Construction Documents

LEED Submittal Phase:

Construction

Related Code or Standard:

None

Responsible Party:

Contractor

MRc3.1: Materials Reuse (1-2 points)

Purpose:

To reuse building products and materials to reduce waste and save original/virgin materials. Also to reduce the environmental impact related to processing and extracting original materials.

Credit Path:

A total of 5%, 10%, or more (based on cost) of the materials used should be reused, refurbished, or salvaged materials. The approved materials can be from off-site sources.

You can only include permanently installed materials for your job. You need to exclude furniture and furnishings.

% of Reused Materials	Points
5%	1
10%	2

Considerations and Benefits:

1) **Environmental:** Materials Reuse reduces the needs for landfills, saves disposal fees, and costs for extracting raw materials, transportation, and manufacturing. It can divert construction waste, preserve open space, and reduce waste and emissions. Its impact can be significant since buildings use a large percentage of natural resources: 25% of virgin wood, 40% of sands and gravels, etc.

2) **Financial:** Materials Reuse may increase the complexity of the project and add extra labor cost, but it can also significantly decrease the material cost. Overall, the owner will probably save money.

   Recyclable materials have different values. Overall, it is cost-effective to recycle cardboard, metals, concrete, and asphalt. Saving disposal fees is also an important incentive.

Submittals:

1) A list of salvaged and reused materials and their related costs. Use construction costs for materials per CSI MasterFormat™ 2004 Edition Divisions 03-10, 31 (section 31.60.00 Foundations) and 32 (section 32.10.00 Paving, 32.30.00 Site Improvements, and 32.90.00 Planting).

OR

2) A list of actual materials costs, not including equipment and labor.

Synergies:

MRc3.1 is related to the following credits:

- MRc1: Building Reuse
- MRc2: Construction Waste Management
- MRc3.2: Materials Reuse: Furniture and Furnishing

Remanufactured materials cannot count towards MRc3.1 because they are NOT considered reuse materials, but they can contribute to the following credits:

- MRc2: Construction Waste Management
- MRc4: Recycled Content

Material costs for MRc3.1 shall be consistent with those used in the following credits:
- MRc4: Recycled Content
- MRc5: Regional Materials
- MRc6: Rapidly Renewable Materials

Possible Strategies and Technologies:

1) Find potential material suppliers and take advantage of every opportunity to incorporate reused materials.
2) Reuse building materials such as doors and frames, columns and beams, paneling, flooring, decorative items, and bricks, etc.
3) Exclude repaired and recycled materials, electrical, mechanical and plumbing items, and elevators and equipment or other specialty items in your calculation.
4) The material cost is typically 45% of the total cost of the building.
5) Furniture and Furnishing are included in MRc3.2, and should NOT be included in MRc3.1
6) Refurbished materials, such as a wood door converted into a computer desk, can be included in either MRc3.1 or MRc3.2, but NOT both.

Extra Credit (Exemplary Performance):

You can get one innovation point for MRc3.1 if 15% of the building materials (based on cost) are reused, refurbished, or salvaged.

Project Phase:

Construction Documents

LEED Submittal Phase:

Construction

Related Code or Standard:

None

Responsible Party:

Contractor and Architect

MRc3.2: Materials Reuse: Furniture and Furnishings (1 point)

Purpose:

To reuse building products and materials to reduce waste and save original/virgin materials, also to reduce the environmental impact related to processing and extracting original materials.

Credit Path:

A total of 30% or more (based on cost) of the furniture and furnishings used should be reused, refurbished, or salvaged. They can be from off-site sources.

Salvage Rate (%) = (Replacement Value of Reused Furniture and Furnishings/Total Value of Reused Furniture and Furnishings) x 100

Considerations and Benefits:

1) **Environmental:** Reusing furniture and furnishings can reduce the needs for landfills, save disposal fees, and costs for extracting raw materials, transportation, and manufacturing, as well as preserve historical value. It can divert construction waste, preserve open space, as well as reduce waste and emissions.

2) **Financial:** Reusing furniture and furnishings can save a substantial amount of money.

Submittals:

1) Keep record of the actual cost of reused, refurbished, or salvaged furniture and furnishings.

Synergies:

MRc3.2 is related to the following credits:

- MRc2: Construction Waste Management
- MRc3.1: Materials Reuse: 5% and 10%

Possible Strategies and Technologies:

1) The project team can buy used furniture from suppliers.
2) Furniture and Furnishing are included in MRc3.2, and should NOT be included for MRc3.1
3) The percentage is based on the replacement value of furniture and furnishings in CSI MasterFormat™ Division 12.

Extra Credit (Exemplary Performance):

The project team get one innovation point if a total of 60% or more (based on cost) of the furniture and furnishings are reused, refurbished, or salvaged.

Project Phase:

Design Development

LEED Submittal Phase:

Construction

Related Code or Standard:

None

Responsible Party:

Contractor and Architect

MRc4: Recycled Content: post-consumer + ½ pre-consumer (1-2 points)

Purpose:

To encourage the use of recycled materials in buildings, and reduce demand on virgin materials and the related environmental impact.

Credit Path:

At least 10% or 20% (based on cost) of the total value of the materials used in the project should equal the sum of post-consumer plus ½ of the pre-consumer content.

% of Recycled Content	Points
10%	1
20%	2

You can use the weight of the recycled material divided by the total weight of the assembly to obtain the percentage (or fraction), and then use the percentage multiplied by the cost of the assembly to get the value of the recycled materials.

Do not include electrical, mechanical or plumbing items, elevator equipment, or other specialty items in you calculation. You can only include materials permanently installed in your job. You can include furniture only if it is consistently included in MRc3 through MRc7.

You can define recycled materials per the International Organization of Standards document, *ISO 14021-Environmental Labels and Declarations, Self-declared Environmental Claims (Type II environmental labeling).*

Post-consumer material refers to waste created by institutional, industrial, or commercial facilities or households that can no longer be used for its intended purpose.

Pre-consumer material refers to waste diverted from the waste flow in the process of manufacturing. You cannot include materials (like regrind, rework, etc.) that can be reclaimed within the same process that they were generated.

Considerations and Benefits:

1) **Environmental:** Products with recycled content reduce the needs for landfills, save disposal fees, and reduce costs for extracting raw materials, transportation, and manufacturing. They can divert construction waste, preserve open space, and reduce waste and emissions.

2) **Financial:** Many products can be made from recycled content. They have little or no extra costs.

Submittals:

1) A list of manufacturers' names, product names, costs, and the percentage of pre- and post-consumer content
2) Manufacturers' letters or cutsheets indicating the listed products' recycled content

Synergies:

MRc4 is related to the following credits:

- MRc2: Construction Waste Management
- MRc3: Materials Reuse

Look for recycled materials that are remanufactured locally and use local waste products for potential synergies with MRc5: Regional Materials.

Some recycled-content materials like synthetic products (rubber, plastic, and polyester) may have problematic emissions. Make sure to coordinate all recycled-content materials with IEQc4: Low-Emitting Materials

Material costs for MRc4 shall be consistent with those used in the following credits:

- MRc3: Materials Reuse
- MRc5: Regional Materials
- MRc6: Rapidly Renewable Materials

Possible Strategies and Technologies:

1) Set a goal for recycled content, find potential material suppliers, and take advantage of every opportunity to recycle materials.
2) Make sure the recycled materials are installed.
3) Consider environmental, performance, and economic factors when choosing materials.

Extra Credit (Exemplary Performance):

You can get one innovation point for MRc4 if 30% of the materials used are recycled.

Project Phase:

Schematic Design

LEED Submittal Phase:

Construction

Related Code or Standard:

International Organization for Standardization (ISO): International Standard ISO 14021-1999, Environmental Labels and Declarations, Self-Declared Environmental Claims (Type II Environmental Labeling)

Responsible Party:

Contractor and Architect

MRc5: Regional Materials (1-2 points)

Purpose:

To encourage the use of materials that are manufactured or extracted within the immediate region of the building; to reduce the environmental impact generated from transportation; and to support the local economy.

Credit Paths:

1st Choice: (1 point)

At least 20% (based on cost) of the total value of new construction and Division 12 (Furniture) products or materials used in the project should be manufactured within 500 miles of the job site. If only a portion of a building product or material is generated locally, include only that portion (by percentage of weight) for this credit.

2nd Choice: (2 points)

Meet the criteria above,

AND

At least 10% (based on cost) of the total value of construction and Division 12 (Furniture) products or materials used in the project should be recovered, harvested, extracted, AND manufactured within 500 miles of the job site. If only a portion of a building product or material is generated locally, include only that portion (by percentage of weight) for this credit.

Do not include electrical, mechanical and plumbing items, and elevators and equipment, or other specialty items in you calculation. You can only include permanently installed materials for your job.

Considerations and Benefits:

1) **Environmental:** Using regional materials reduces transportation demands and related emissions.

2) **Financial:** Using regional materials reduces transportation costs, and supports the local economy.

Submittals:

1) A list of product purchases extracted, harvested, or manufactured regionally.
2) Paperwork showing manufacturers' names, product names, costs, the distance between the extraction site and the project, and the distance between the manufacturer and the project
3) Manufacturers' letters or cutsheets indicating the products' manufacture within a 500 miles radius of the site
4) For the 2nd choice, include manufacturers' letters or cutsheets indicating the materials' extraction within a 500 miles radius of the site

Synergies:

MRc5 may affect the following credits:

- MRc3: Materials Reuse
- MRc4: Recycled Content
- MRc6: Rapidly Renewable Materials

Material costs for MRc5 shall be consistent with those used in the following credits:

- MRc3: Materials Reuse
- MRc4: Recycled Content
- MRc6: Rapidly Renewable Materials

Possible Strategies and Technologies:

1) Set a goal for local materials, find potential material suppliers, and take advantage of every opportunity to use local materials.
2) Assure the local materials are installed.
3) Take into account environmental, performance, and economic factors when choosing materials.

Extra Credit (Exemplary Performance):

You can get one innovation point for MRc5 if 20% of the materials are recovered, harvested, extracted, AND manufactured within 500 miles of the job site.

Project Phase:

Schematic Design

LEED Submittal Phase:

Construction

Related Code or Standard:

None

Responsible Party:

Contractor and Architect

MRc6: Rapidly Renewable Materials (1 point)

Purpose:

To use rapidly renewable materials to replace long-cycle renewable and limited raw materials.

Credit Path for NC and Schools:

At least 5% (based on cost) of total value of the new construction and Division 12 (Furniture and Furnishings) products and materials for your job should be rapidly renewable materials (produced from plants that can be generated within a 10-year cycle).

Considerations and Benefits:

1) **Environmental:** Rapidly renewable materials are replenished faster than traditional materials, and they can help to reduce the negative environmental impacts of construction. For example, using wheat in composite panels can substantially reduce the use of wood and the related negative environmental impacts.

2) **Financial:** Rapidly renewable materials can save land, and provide faster payback.

Submittals:

1) A list of rapidly renewable product purchases
2) Paperwork showing the manufacturers' names, product names, costs, percentage of rapidly materials (by weight) in each product, and compliance value
3) Manufacturers' letters or cutsheets confirming rapidly materials

Synergies:

Rapidly renewable materials may come from a distant location and may affect the following credit:

- MRc5: Regional Materials

Specify materials that do not release irritating or harmful chemicals, such as VOCs from solvents and paints. Coordinate MRc6 with the following credit:

- IEQc4: Low-Emitting Materials

Material costs for MRc6 shall be consistent with those used in the following credits:

- MRc3: Materials Reuse
- MRc4: Recycled Content
- MRc5: Regional Materials

Possible Strategies and Technologies:

1) Set a goal for the use of rapidly renewable materials, find potential material suppliers, and take advantage of every opportunity to use rapidly renewable materials.
2) Assure the rapidly renewable materials are installed.
3) Rapidly renewable materials include wool, bamboo, cotton, wheatboard, cork, strawboard, cotton insulation, agrifiber, natural rubber, linseed (linoleum), etc.

Extra Credit (Exemplary Performance):

You can get one innovation point for MRc6 if 10% of the materials are rapidly renewable materials.

Project Phase:

Schematic Design

LEED Submittal Phase:

Construction

Related Code or Standard:

None

Responsible Party:

Contractor and Architect

MRc7: Certified Wood (1 point)

Purpose:

To encourage environmentally friendly forest management, ban the use of pesticides, and to protect the health and safety of forest workers.

Credit Path for NC, Schools, and CS:

At least 50% (based on cost) of new wood building components should be certified per the **Forest Stewardship Council's (FSC)** principles and criteria. These wood components include general dimensional framing, structural framing, sub-flooring, flooring, finishes, wood doors, etc.

Only include permanently installed materials. You can include material value for Division 12, Furniture.

Exception to this rule: You can include purchased wood products for temporary use like bracing, formwork, scaffolding, guardrail, and sidewalk protection, etc. If you decide to include these wood products, you need to include ALL such materials. If you purchase these products for multiple projects, you can ONLY include them for one project of your choice.

Considerations and Benefits:

1) **Environmental:** The **Forest Stewardship Council's (FSC)** principles and criteria can alleviate the negative environmental impacts associated with irresponsible forest practices, such as forest destruction, waste generation, water and air pollution, wildlife habitat loss, erosion, and waterways sedimentation.

2) **Financial:** FSC-certified wood products' costs are equal to or higher than conventional wood products. Their prices are expected to become more competitive as more and more consumers demand them. The **Forest Stewardship Council's (FSC)** principles and criteria can stabilize the local economy and preserve our limited forest resources for future generations.

Submittals:

1) A list identifying the percentage of certified wood in each purchase
2) Record of the certified wood purchase and related **Chain of Custody (COC)** paperwork
3) Vendor invoices for certified wood products
4) The CSI MasterFormat 2004 Division 3-10 and Division 12 (Furniture and Furnishings) determining the net construction material cost.

Synergies:

Materials used for MRc7 may also qualify for:

- IEQc4.4: Low-Emitting Materials: Composite Wood and Agrifiber Products
- MRc5: Regional Materials

Possible Strategies and Technologies:

1) Set a goal for the use of FSC certified wood products, find potential material suppliers, and take advantage of every opportunity to use rapidly renewable materials.
2) Assure the FSC certified wood products are installed and quantified in terms of percentage.
3) There are two types of certification: a. Chain of Custody (COC) for the suppliers; and b. All trades' forest management.
4) Suppliers need to obtain a certification number, but the contractor does NOT need to get one.
5) If you need to get a certification number for part of the product, then you need to get certification for the whole product.
6) Using lower grades of wood can substantially reduce the demand on forests because top-grade timber is very limited. For example, the project team may be able to use Architectural Wood Work Institute Grades 2 or 3 instead of Grade 1 for veneer or lumber.

Extra Credit (Exemplary Performance):

The project team can get one innovation point if 95% or more of the wood building components are certified per **Forest Stewardship Council's (FSC)** principles and criteria.

Project Phase:
Schematic Design

LEED Submittal Phase:
Construction

Related Code or Standard:
Forest Stewardship Council's (FSC) principles and criteria:
http://www.fscus.org

Responsible Party:
Contractor and Architect

MR Summary and Mnemonics:

MR Credit Name	Extra Credit	Responsible Party
MRp1: **S**torage and Collection of Recyclables (Mandatory, 0 points)	0	Architect, Owner and tenant
MRc1.1: Tenant Space: Long-Term Commitment	0	Architect and Owner
MRc1.2: **B**uilding Reuse: Maintain Interior Nonstructural Components (1-2 points)	1 extra ID point for keeping 80% of the existing nonstructural elements like doors, floor coverings, ceiling systems, and interior walls (based on surface area).	Architect, Contractor, and Owner
MRc2: **C**onstruction Waste Management (1-2 points)	1 extra ID point for MRc2 by salvaging and/or recycling a minimum of 95% of total construction debris.	Contractor
MRc3.1: **M**aterials Reuse (1-2 points)	1 extra ID point for MRc3.1 if 15% of the building materials (based on cost) are reused, refurbished, or salvaged.	Contractor and Architect
MRc3.2: Materials Reuse: Furniture and Furnishings (1 point)	1 extra ID point if a total of 60% or more (based on cost) of the furniture and furnishings are reused, refurbished, or salvaged.	
MRc4: **R**ecycled Content: post-consumer + ½ pre-consumer (1-2 points)	1 extra ID point for MRc4 if 30% of the materials used are recycled.	Contractor and Architect
MRc5: **R**egional Materials (1-2 points)	1 extra ID point for MRc5 if 20% of the materials are recovered, harvested, extracted, AND manufactured within 500 miles of the job site.	Contractor and Architect
MRc6: **R**apidly Renewable Materials (1 point)	1 extra ID point for MRc6 if 10% of the materials are rapidly renewable materials.	Contractor and Architect
MRc7: **C**ertified Wood (1 point)	1 extra ID point if 95% of the wood building components are certified per Forest Stewardship Council's (FSC) principles and criteria.	Contractor and Architect

Mnemonics:
SBC
MR Regan Ray Carter (See bold and underlined letters in credit names on table above also). Of course, you can create your own **mnemonics** to help you memorize them.

E. Indoor Environmental Quality (IEQ)

The average American spends about 90% of his time indoors, so indoor environmental quality is very important for quality of life, well-being, and productivity.

Overall purpose:

1) Design and install proper systems to achieve a quality indoor environment
2) Reduce, manage, and eliminate contaminants
3) Minimum IAQ Performance
4) Environmental Tobacco Smoke (ETS) Control
5) Outdoor Air Delivery Monitoring
6) Increased Ventilation
7) Construction IAQ Management Plan
8) Low-Emitting Materials
9) Indoor Chemical and Pollutant Source Control
10) Controllability of Systems
11) Thermal Comfort
12) Daylight and Views

Mnemonics:
I Call Mike Evans.
Oh! Ian Catches LIST D (See underlined letters in credit names on table above also.)

Core concepts:

1) Improve indoor air quality
   - Improve building ventilation
   - Choose proper materials
   - Reduce, manage, and eliminate contaminants
   - Advocate green construction practices and green building operation
2) Improve indoor environmental quality
   - Thermal comfort control
   - Daylight and views
   - Considering acoustics

Recognition, regulation and incentives:

1) Guidance is available from private and public sector organizations such as:
   - EPA
   - Department of Labor
   - ASHRAE

Overall strategies and technologies:

1) Choose low-emitting materials, interior finishes, furniture, etc.
2) Reduce, manage, and eliminate contaminants like certain cleaning products, tobacco smoke, and radon.
3) Advocate green construction practices like **Best Management Practices (BMPs)**, outdoor air introduction, green cleaning, and proper handling of exhaust systems.
4) Thermal comfort control: user control and feedback, operations and maintenance/ management.
5) Provide daylight/views: north-facing skylights, interior light (reflecting) shelves, interior and exterior permanent shading devices, automatic photocell-based controls and high performance glazing to maximize daylight for interior spaces while avoiding high-contrast conditions.
6) Use light fixtures with sensors and dimming controls.

Specific Technical Information:

IEQp1: Minimum IAQ Performance (Mandatory, 0 points)

Purpose:

To improve indoor air quality (IAQ) as well as the well being and comfort of building occupants by establishing minimum IAQ performance.

Credit Paths:

The building should meet Section 4 through 7 of ASHRAE 62.1-2007, Ventilation for Acceptable Indoor Air Quality (with errata but without addenda).

1) **Active (Mechanical)** Ventilation: The mechanical ventilation systems shall follow Ventilation Rate Procedure or local codes, if they are more restrictive.

   Maintain or modify existing mechanical ventilation systems to meet the outdoor ventilation rate per ASHRAE 62.1-2007, Ventilation for Acceptable Indoor Air Quality (with errata but without addenda).

   If the project team cannot achieve the above requirements, describe the constraints of existing space and system, and achieve the maximum levels possible (based on an engineering assessment) toward meeting the above criteria, but the project still needs to achieve at least 10 cfm per person. The project team still needs to meet all other criteria.

OR

2) **Passive (Natural)** Ventilation: Your building should follow ASHRAE 62.1-2007 paragraph 5.1 (with errata but without addenda), if it is naturally ventilated.

   For both paths:

   Regardless of the ventilation mode, your building should follow ASHRAE 62.1-2007, Chapter 6 (with errata but without addenda).

Considerations and Benefits:

1) **Environmental:** Achieving minimum IAQ improves occupants' well-being, comfort, and productivity. The key methods to achieve this are preventing the outdoor contaminants from entering the building, limiting indoor contaminants, and maintaining or modifying existing mechanical ventilation systems to meet the outdoor ventilation rate per ASHRAE 62.1-2007.

2) **Financial:** A project should not incur any extra costs for meeting ASHRAE 62.1-2007 since it is already a standard practice in many areas. If a remodel project requires extra energy cost for improving IAQ, the project team may use CO_2 monitors, controls and economizers, energy recovery ventilation, and demanded controlled ventilation to alleviate these costs.

   Poor IAQ causes sick building syndromes and poor productivity. Compared with labor cost, any extra cost for improving IAQ is negligible.

Submittals:

1) Calculations to confirm compliance with ASHRAE 62.1-2007

Synergies:

You can dilute contaminant concentration and solve some IAQ problems, but this measure may increase energy use and affect indoor thermal comfort. There are ways to minimize energy loss and improve IAQ as described in the following prerequisite and credits:

- EAp1: Fundamental Commissioning of the Building Energy Systems
- EAc2: Enhanced Commissioning
- EAc3: Measurement and Verification

Existing site contamination, heavy traffic, and dense neighborhoods may have a negative impact on the quality of outside air for ventilation. Refer to the following credits:

- SSc1, Ch2F1 (Choice Two, Feature 1): Brownfield Redevelopment
- SSc3: Alternative Transportation

Some materials may have negative effects on IAQ. Try to specify furnishings and materials free of irritating and harmful chemicals such as VOCs from solvents and paints. Refer to the following prerequisite and credits:

- IEQp2: Environmental Tobacco Smoke (ETS) Control
- IEQc4: Low-Emitting Materials
- IEQc5: Indoor Chemical and Pollutant Source Control

Possible Strategies and Technologies:

a. Design the building's ventilation system to meet or exceed ASHRAE requirements. Optimize and balance occupant health with energy efficiency.
b. Refer to the ASHRAE 62 User's Manual for more information.
c. For **Passive (Natural)** Ventilation, the openings for ventilation shall be at least 4% of the floor area.
d. IEQp1 is NOT limited to the project's scope of work since the tenant space may be served by a central HVAC system outside of the tenant space.

e. Evaluate the potential space for possibility of meeting IEQp1 before leasing or purchasing the property.

Extra Credit (Exemplary Performance):
None

Project Phase:
Schematic Design

LEED Submittal Phase:
Design

Related Code or Standard:
ANSI/ASHRAE 62.1-2007 (with errata but without addenda)

Responsible Party:
MEP Engineer

IEQp2: Environmental Tobacco Smoke (ETS) Control (Mandatory, 0 points)

Purpose:
To minimize or prevent the impact of Environmental Tobacco Smoke (ETS) on ventilation air distribution systems, indoor surfaces, and occupants.

Credit Paths:

For Non-Residential Projects:

1st Path:

Choose a tenant space in a building that:

a. Places any designated exterior smoking area a minimum of 25 feet away from operable windows, entries, and outdoor air intakes. Use signage to designate smoking areas, and prohibit smoking in certain areas, or prohibit smoking on the entire property.
b. Prohibits indoor smoking.

OR

2nd Path:

Choose a tenant space in a building that:

a. Places any designated exterior smoking area a minimum of 25 feet away from operable windows, entries, and outdoor air intakes. Use signage to designate smoking areas, and prohibit smoking in certain areas, or prohibit smoking on the entire property.
b. Prohibits indoor smoking, except in designated smoking areas/rooms.
c. Places smoking rooms in a location to effectively contain, capture, and remove ETS. The smoking room should be exhausted directly to the outdoors and enclosed with deck-to-deck impermeable partitions. The ETS-containing air shall never be re-circulated to non-smoking areas. When the smoking room doors are closed, the smoking room should have negative pressure of a minimum of 1 Pa (0.004 inches of water gauge), and an average of 5 Pa (0.02 inches of water gauge).
d. Has conducted a 15-minute measurement with the smoking room doors closed. There should be

1 measurement of at least every 10 seconds of the pressure between the smoking rooms, each adjacent vertical chase, and each adjacent area. The tests should be done for the worst case conditions.

For Residential Projects:

3rd Path:

a. Place any designated exterior smoking area a minimum of 25 feet away from operable windows, entries, and outdoor air intakes to common areas. Use signage to designate smoking areas, and prohibit smoking in certain areas, or prohibit smoking on the entire property.
b. Prohibit smoking in the building's common areas.
c. Sealing vertical chases, penetrations in floors, walls, and ceilings to minimize ETS transfer between units.
d. For residential units, install weather-striping on all operable windows, exterior doors, and doors leading to common hallways.
e. The residential units adjacent to the common pathways do not need to have weather-striping if the common hallways are pressurized per Path 2 above.
f. The sealing of residential units shall pass a blower door test per ANSI/ASTM-E779-03, **Standard Test Method** for Determining Air Leakage Rate by Fan Pressure, and the **Progressive Sampling Method** per Chapter 7 (Home Energy Rating Systems (HERS) Required Verification and Diagnostic Testing) of the California Low Rise Residential Alternative Calculation Method Approval Manual (http://www.energy.ca.gov/). Residential units should have less than 1.25 square inches of leakage area per 100 square feet of enclosed area (the total of floor, wall, and ceiling areas).

Considerations and Benefits:

1) **Environmental:** Smoking or secondhand smoke causes many diseases and has severe impacts on human health. IEQp2 is intended to reduce or eliminate these impacts. Providing isolated and sealed indoor smoking spaces may make the building larger, but can minimize or prevent the impact of Environmental Tobacco Smoke (ETS) on ventilation air distribution systems, indoor surfaces, and occupants.

2) **Financial:** IEQp2 can have higher design and construction costs because of the isolated and sealed indoor smoking spaces, and higher maintenance and operation costs for providing a separate outdoor smoking area, but it can substantially reduce insurance and health care costs, increase the lifespan of the interior furnishings and furniture, and increase the desirability of the tenant spaces.

Submittals:

1) An ETS policy describing non-smoking areas
2) Renderings, site plans, or other documentation showing how the smoking policy is implemented
3) A record of testing data for interior smoking rooms showing that there is no cross contamination

Synergies:

A separate ventilation system for smoking areas adds cost to commissioning, energy, and M&V efforts. Refer to the following prerequisite and credits:

- EAp1: Fundamental Commissioning of the Building Energy Systems
- EAc1: Optimize Energy Performance
- EAc2: Enhanced Commissioning
- EAc3: Measurement and Verification

Both outdoor and indoor smoking affects IAQ performance. IEQp2 is related to the following prerequisite and credits:

- IEQp1: Minimum IAQ Performance
- IEQc1: Outdoor Air Delivery Monitoring
- IEQc2: Increased Ventilation

You may want to address smoking along with other related air pollutants. Refer to the following credits:

- IEQc4: Low-Emitting Materials
- IEQc5: Indoor Chemical and Pollutant Source Control Products

Possible Strategies and Technologies:

Choose a building that meets the following criteria:

1) For commercial buildings, effectively design and control ventilation of smoking rooms or prohibit smoking.
2) For residential projects, prohibit smoking in common areas, and design and construct building systems and envelope to minimize ETS transfer between the dwelling units and common areas.

Extra Credit (Exemplary Performance):

None

Project Phase:

Pre-Design

LEED Submittal Phase:

Design

Related Code or Standard:

1) ANSI/ASTM-E779-03, Standard Test Method for Determining Air Leakage Rate by Fan Pressure
2) Chapter 7 (Home Energy Rating Systems (HERS) Required Verification and Diagnostic Testing) of the California Low Rise Residential Alternative Calculation Method Approval Manual

Responsible Party:

LEED AP, Owner, and MEP Engineer

IEQc1: Outdoor Air Delivery Monitoring (1 Point)

Purpose:

The intent is to improve occupant well-being and comfort by providing ways to monitor ventilation systems.

Credit Paths:

Install permanent monitoring systems to check the performance of the ventilation system. Set an alarm to go off when the ventilation system performance (CO_2 levels or air flow values) deviates 10% or more from the set point. The alarm can be an audible or visual alarm to building occupants or a building automation system alarm to the building operator.

1) **For Naturally Ventilated Spaces**
   Monitor CO_2 concentrations in all spaces that are naturally ventilated. Place CO_2 monitoring equipment between 3' and 6' above the floor in the rooms. You can use one CO_2 sensor for multiple non-densely spaces if you use **passive stack(s)** or other measures to induce air flow through these spaces simultaneously and equally without the help of occupants.

2) **For Mechanically Ventilated Spaces**
   Monitor CO_2 concentrations in all densely occupied spaces, i.e., spaces with a design occupant density greater than or equal to 25 persons per 1000 s.f. Place CO_2 monitoring equipment between 3' and 6' above the floor in the rooms.

For spaces where 20% or more of the supply air flow serves areas with a design occupant density less than 25 persons per 1,000 s.f., provide a direct outdoor airflow monitoring/measurement device that is able to check the minimum outdoor airflow rate with an accuracy rate of plus or minus 15% of the minimum design outdoor airflow rate per ASHRAE 62.1-2007 (with errata but without addenda).

Considerations and Benefits:

1) **Environmental:** High concentration of CO_2 can lead to increased heart rate, dizziness, and headaches. Monitoring CO_2 concentration is a way to measure the effectiveness of outdoor air ventilation.

2) **Financial:** Ventilation rate and CO_2 monitoring systems increase the initial investment of a project, as well as maintenance and operation costs, but they provide valuable benefits: reduced HVAC energy use and absenteeism, as well as increased productivity.

Submittals:

1) Show CO_2 sensors and air flow monitors on schematics, floor plans, elevations, and mechanical schedules.
2) Use commissioning to monitor the ventilation system energy usage.
3) Check alarm system settings per ANSI/ASHRAE 62.1-2007 for mechanical ventilation systems.
4) Adjust all automated building systems per manufacturer's recommendations. Perform routine check for alarm systems.

Synergies:

EAc1 can alert building users about potential IAQ problems and create opportunities for performance trending. Monitoring capacity enables robust M&V and provides information for commissioning, ensures consistent IAQ, and maximizes energy performance. Refer to the following prerequisite and credits:

- IEQc2: Increased Ventilation

- EAp1: Fundamental Commissioning of the Building Energy Systems
- EAc2: Enhanced Commissioning
- EAc3: Measurement and Verification

You can increase ventilation rates to mitigate indoor CO_2, but the outdoor air quality depends on the CO_2 in the outside air. Heavy traffic, dense neighborhoods, and existing site contamination can increase CO_2 levels and affect the outdoor air quality. Bicycle corridors, public transportation, and other transportation amenities can reduce the number of solo drivers and lower CO_2 levels. Refer to:

- SSc1, Ch2F1: Brownfield Redevelopment
- SSc3: Alternative Transportation

Possible Strategies and Technologies:

1) Install airflow and carbon dioxide monitoring equipment, and then transfer the information to the Building Automation System (BAS) and/or HVAC system to trigger corrective action.
2) If you cannot complete strategy 1, then at minimum use the monitoring equipment to trigger alarms to alert occupants or operators of the poor outdoor air delivery.

Extra Credit (Exemplary Performance):
None

Project Phase:
Schematic Design

LEED Submittal Phase:
Design

Related Code or Standard:
ANSI/ASHRAE 62.1-2007: Ventilation for Acceptable Indoor Air Quality

Responsible Party:
MEP Engineer

IEQc2: Increased Ventilation (1 Point)

Purpose:
To provide increased outdoor air ventilation to improve indoor air quality, occupants' well-being, comfort, and productivity.

Credit Paths:

For Naturally Ventilated Spaces:
Study and follow the flow diagram process in Figure 2.8 of the Chartered Institution of Building Service Engineers (**CIBSE**) Applications Manual 10: 2005, Natural Ventilation in Non-Domestic Buildings, and determine if natural ventilation is an effective choice for your job.

AND

1) Use calculations and diagrams to show that your system meets the following requirements:
Choice 1: CIBSE Applications Manual 10: 2005, Natural Ventilation in Non-Domestic Buildings
Choice 2: CIBSE Applications Manual 13: 2000, Mixed Mode Ventilation

OR

1) Use an analytic, multi-zone, macroscopic model to show that airflow in each room will effectively naturally ventilate and meet the minimum ventilation rates per ASHRAE 62.1-2007 Chapter 6 (with errata but without addenda) for a minimum of 90% of the occupied spaces.

OR

For Mechanically Ventilated Spaces:

2) Increase outdoor air ventilation rates to breathing zones of all occupied spaces by a minimum of 30% above the minimum rates per ASHRAE Standards 62.1-2007. See EQp1 for more information.

Note:

For ALL options, breathing zone means space 2' from exterior walls and 3' to 6' above finished floor (AFF). Use 0.30 CFM per s.f. for outside air.

Considerations and Benefits:

1) **Environmental:** Indoor biological pollutants and chemicals can lead to many health problems. Increasing outdoor air ventilation can improve indoor air quality, occupants' well-being, comfort, and productivity.

2) **Financial:** Increasing outdoor air ventilation can increase the initial investment of a project as well as operation costs, but also provides valuable benefits. Naturally ventilated buildings may cost more and need better insulated windows and walls, but the HVAC and maintenance costs will be lower.

Submittals:

1) Size mechanical equipment to match increase ventilation rates
2) Keep plans and other proper documentation of open areas within the project for naturally ventilated projects

Synergies:

Ventilation, especially mechanical ventilation, requires commissioning, M&V, and building energy performance. A permanent ventilation performance monitoring system can help to achieve and maintain increased ventilation. Refer to the following related prerequisites and credits:

- EAp1: Fundamental Commissioning of the Building Energy Systems
- EAp2: Minimum Energy Performance
- EAc1: Optimize Energy Performance
- EAc2: Enhanced Commissioning
- EAc3: Measurement and Verification

- IEQc1: Outdoor Air Delivery Monitoring

Possible Strategies and Technologies:

You can use LoopDA, Natural Ventilation Sizing Tool, and NIST's CONTAM, Multizone Modeling Software to show airflows in each room analytically.

Use heat recovery if possible to reduce energy consumption caused by higher ventilation rates for mechanically ventilated spaces.

Extra Credit (Exemplary Performance):
None

Project Phase:
Schematic Design

LEED Submittal Phase:
Design

Related Code or Standard:
1) Chartered Institution of Building Service Engineers (CIBSE) Application Manual 10, Natural Ventilation in Non-Domestic Buildings, 2005
2) ASHRAE Standards 62.1-2007, Ventilation for Acceptable Indoor Air Quality

Responsible Party:
Contractor and MEP Engineer

IEQc3.1: Construction IAQ Management Plan: During Construction (1 Point)

Purpose:
To sustain the well-being and comfort of building occupants and construction workers by reducing air pollution caused by the renovation/construction process.

Credit Paths:
Create and implement an IAQ (Indoor Air Quality) Management Plan for the pre-occupant and construction phases:
1) Use filtration media with a Minimum Efficiency Reporting Value (MERV) of **8** at each return air grille per ASHRAE 52.2-1999 (with errata but without addenda) if a permanently installed air handler is used during construction. Change all filtration media immediately before occupancy.
2) Protect installed or stored on-site absorptive materials from moisture damage.
3) At minimum, meet the Control Measures of the Sheet Metal and Air Conditioning National Contractors Association (SMACNA) IAQ Guideline for Occupied Buildings under Construction, 2nd Edition, November 2007, Chapter 3, During Construction.

Considerations and Benefits:
1) **Environmental:** Demolition and construction activities generate many indoor pollutants which have negative impacts on human health. Construction IAQ management plan can improve indoor air quality, occupants' well-being, comfort, and productivity.

2) **Financial:** Isolating the activities involving power equipment and protecting the ventilation system are two of the most effective ways of IAQ management plan. IAQ management plan can increase the initial investment of a project as well as the operation costs, but can also provide valuable benefits. Proper sequencing and early coordination can eliminate or reduce scheduling delays.

Submittals:

1) A written construction IAQ management plan for demolition and construction
2) A detailed photo log and brief description of the construction IAQ management plan practices used during construction

Synergies:

Refer to the following related credits:

- IEQc3.2: Construction IAQ Management Plan: Before Occupancy
- IEQc4: Low-Emitting Materials
- IEQc5: Indoor Chemical and Pollutant Source Control

Possible Strategies and Technologies:

1) Interrupt contamination pathways, control pollutant sources, and protect the HVAC system by creating and implementing an IAQ management plan.
2) Properly sequence material installation to avoid contamination of carpeting, ceiling tiles and gypsum wallboard, insulation, and other absorptive materials. Coordinate with IEQc3.2 and IEQc5; select the proper schedules and specifications for filtration media.
3) Try not to use permanently installed air handlers for temporary cooling/heating during construction. Refer to the related USGBC reference guide for more information.

Extra Credit (Exemplary Performance):

None

Project Phase:

Construction Administration

LEED Submittal Phase:

Construction

Related Code or Standard:

1) ASHRAE 52.2-1999
2) Sheet Metal and Air Conditioning National Contractors Association (SMACNA) IAQ Guideline for Occupied Buildings under Construction, 2nd Edition, November 2007, Chapter 3.

Responsible Party:

Contractor and MEP Engineer

IEQc3.2: Construction IAQ Management Plan: Before Occupancy (1 Point)

Purpose:

To sustain the well-being and comfort of building occupants and construction workers by reducing air pollution caused by the renovation/construction process.

Credit Paths:

Create and implement an IAQ (Indoor Air Quality) Management Plan after the installation of all finishes and the cleaning of the entire building but before occupancy:

1st Path (Flush-out):

After construction and all interior finishes are completed but before occupancy, conduct a flush-out by keeping the relative humidity less than or equal to 60%, and a minimum of 60 degrees Fahrenheit internal temperature, and provide a total air volume of 14,000 cubic feet (c.f.) of outdoor air per s.f. of floor area.

OR

If a space will be used before flush-out, supply at least 3,500 c.f. of outdoor air per s.f. of floor area before using the space. Once you occupy the space, you need to ventilate it at least at a rate of 0.30cfm/s.f. of outside air or meet the minimum requirement per EQp1 if it is more restrictive. Ventilate the space for at least three hours before you use the space and continue to ventilate it while you use the space. You need to maintain these until you supply a total of 14,000 c.f./s.f. of outside air.

Note: Start the flush-out AFTER completion of cleaning, the punch list, and final test of the control systems.

OR

2nd Path (Air Testing):

a. Per the U.S. Environmental Protection Agency (EPA) Compendium of Methods for the Determination of Air Pollutants in Indoor Air, conduct a baseline IAQ test after construction, but before occupancy. Refer to the related USGBC Reference Guide for more information.

   Also make sure that your spaces have not exceeded the maximum contaminant concentrations listed per the related USGBC reference guide.

b. If any of the limits above is exceeded at a sampling point, you need to do additional flush-outs until a retest showing the above limit is met. If you need to retest non-complying areas, take samples from the same locations as in the first test.

c. Conduct air sampling tests per the following criteria:
   1) Before occupancy, simulate the normal occupancy conditions, i.e., do the test during normal hours, start the ventilation at normal hours, and run it at minimum outside air flow rate for the occupied mode for the entire length of the testing.
   2) All permanent interior finishes (carpet and acoustic ceiling tiles, paint, doors, millwork, etc.) should have been completed. You are required to complete installation of movable furnishings.
   3) The number of ventilation systems and size of the building dictate the number of sampling locations. You should have at least one sampling point for each continuous floor area or for every 25,000 s.f. if a larger number is required. Also, include the areas with the worst ventilation and with the best ventilation.
   4) Take the samples between 3' to 6' high (the breathing zone of a person) and for at least 4 hours.

Considerations and Benefits:

See IEQc3.1

Submittals:

1) A written construction IAQ management plan
2) Documentation for occupancy, dates, outdoor air monitoring rates, humidity, internal temperature, and special considerations for flush-out procedure
3) A copy of the testing report, confirmation of the correct unit of measurement used, and inclusion of all contaminants for IAQ testing

Synergies:

The same construction IAQ management plan often includes best practices during construction and after construction but before occupancy. Refer to the following related credit:

- IEQc3.1: Construction IAQ Management Plan: During Construction

Building external moisture barrier and filtration materials can affect IAQ and related testing. Refer to the following related credit:

- IEQc4: Low-Emitting Materials
- IEQc5: Indoor Chemical and Pollutant Source Control

Outdoor air can be introduced to dilute indoor air contaminants. Refer to the following related prerequisite and credit:

- IEQp1: Minimum IAQ Performance
- IEQc2: Increased Ventilation

Possible Strategies and Technologies:

1) Conduct a flush-out if you do not need to use the building right away.
2) If you do need to use the building right away, you can do tests for air contaminant levels, but these tests can be more expensive than a flush-out. Coordinate with IEQc3.1 and IEQc5 for filtration media.

Extra Credit (Exemplary Performance):

None

Project Phase:

Construction Administration

LEED Submittal Phase:

Construction

Related Code or Standard:

U.S. Environmental Protection Agency (EPA) Compendium of Methods for the Determination of Air Pollutants in Indoor Air

Responsible Party:

Contractor and MEP Engineer

IEQc4.1: Low-Emitting Materials: Adhesive and Sealants (1 Point)

Purpose:

To reduce the number of harmful, irritating, and/or odorous indoor air contaminants; and to provide comfort and protect the well-being of building users and workers who install the materials.

Credit Path:

Use building interior sealants and adhesives (applied on-site and placed inside of the weatherproofing system) per the following standards:

a. Sealants, Sealant Primers and Adhesive: SCAQMD (South Coast Air Quality Management District) Rule #1168, VOC limits with an effective date of 7/1/2005, and rule amendment dated 1/7/2005.

   Refer to the table in the related USGBC reference guide for VOC limits.

   Note: VOC means "Volatile Organic Compound." VOCs are a major contributing factor to ozone depletion and a common air pollutant which has been proven to be a public health hazard.

b. Aerosol Adhesive: Green Seal Standard for Commercial Adhesives GS-36 requirements with an effective date of 10/19/2000. Refer to the related USGBC reference guide for the table of VOC limits.

Considerations and Benefits:

1) **Environmental:** Many building products contain harmful compounds which have negative impacts on human health. The use of low-emitting materials can improve indoor air quality, occupants' well-being, comfort, and productivity.

2) **Financial:** Low-emitting materials can increase the initial investment of a project as well as operation costs, but also provides valuable benefits. As the green building market matures, their prices are expected to drop.

Submittals:

1) A list of interior sealants, sealant primers, and aerosol adhesive used, including product name, manufacturer's name, specific VOC data (g/L, less water), and the related allowable VOC from the reference standard
2) If the VOC budget approach is used, track and tabulate each item and the corresponding VOC amount.

Synergies:

Refer to the following related prerequisite and credits:

- IEQp2: Environmental Tobacco Smoke (ETS) Control

- IEQc3.1: Construction IAQ Management Plan: During Construction
- IEQc3.2: Construction IAQ Management Plan: Before Occupancy

- IEQc4.2: Low-Emitting Materials: Paints and Coatings
- IEQc4.3: Low-Emitting Materials: Flooring Systems

- IEQc4.4: Low-Emitting Materials: Composite Wood and Agrifiber Products
- IEQc4.5: Low-Emitting Materials: Systems Furnishings and Seating

- IEQc5: Indoor Chemical and Pollutant Source Control

Possible Strategies and Technologies:

1) Make sure that you list the VOC limits, or at least refer to the related USGBC reference guide for VOC limits in each specification section.
2) Request cutsheets, MSD, and certification as part of the submittals and review them for VOC limit compliance.
3) Mention VOC limits in the pre-bid meeting and bid-award meeting.

Extra Credit (Exemplary Performance):

None

Project Phase:

Design Development

LEED Submittal Phase:

Construction

Related Code or Standard:

1) SCAQMD (South Coast Air Quality Management District) Rule #1168, VOC limits, effective 1/7/2005
2) Green Seal Standard for Commercial Adhesives GS-36 requirements, effective 10/19/2000
3) California Department of Health Service (DHS) Standard Practice for the Testing of Volatile Organic Emissions from Various Sources Using Small-Scale Environmental Chambers, including 2004 Addenda

Responsible Party:

Contractor and Architect

IEQc4.2: Low-Emitting Materials: Paints and Coatings (1 Point)

Purpose:

To reduce the number of harmful, irritating, and/or odorous indoor air contaminants. Also to provide comfort and protect the well-being of building users as well as the workers who install the materials.

Credit Path for NC and CS:

Use building interior paints and coatings (applied on-site and placed inside of the weatherproofing system) per the following standards:

Architectural coatings, paints, and primers used on interior ceilings and walls shall not exceed the VOC content limits per Green Seal Standard GS-11, Paints, 1st Edition, 5/20/1993.
Flats: 50 g/L
Non-Flats: 150 g/L

Anti-rust and anti-corrosive paints used on interior ferrous metal substrates shall not exceed the VOC content limits of 250 g/L per Green Seal Standard GS-03, Anti-Corrosive Paints, 2nd Edition, 1/7/1997.

Floor coatings, clear wood finishes, primers, stains, and shellacs used for interior items shall not exceed the VOC content limits per SCAQMD (South Coast Air Quality Management District) Rule 1113, Architectural Coatings, rules with an effective date of 1/1/2004.
Clear wood finishes: varnish 350 g/L; lacquer 550 g/L
Floor coatings: 100 g/L
Sealers: waterproofing sealers 250 g/L; sanding sealers 275 g/L; all other sealers 200 g/L
Shellacs: clear 730 g/L; pigmented 550 g/L
Stains: 250 g/L

Considerations and Benefits:
See IEQc4.1

Submittals:
1) A list of interior coating products and paints used, including product names, manufacturers' name, specific VOC data (g/L, less water), and the related allowable VOC from the reference standard
2) If the VOC budget approach is employed, track the amount used

Synergies:
Same list of prerequisites and credits as EAc4.1, except replace EAc4.2 with EAc4.1

Possible Strategies and Technologies:
Same as IEQc4.1

Extra Credit (Exemplary Performance):
None

Project Phase:
Design Development

LEED Submittal Phase:
Construction

Related Code or Standard:
1) SCAQMD (South Coast Air Quality Management District) Rule 1113, Architectural Coatings, rules with an effective date of 1/1/2004.
2) Green Seal Standard GS-03 and GS-11
3) California Department of Health Service (DHS) Standard Practice for the Testing of Volatile Organic Emissions from Various Sources Using Small-Scale Environmental Chambers, including 2004 Addenda

Responsible Party:
Contractor and Architect

IEQc4.3: Low-Emitting Materials: Carpet Systems (1 Point)

Purpose:

To reduce the number of harmful, irritating, and/or odorous indoor air contaminants. Also to provide comfort and protect the well-being of the building users as well as the workers who install the materials.

Credit Paths:

1st Path:

All flooring shall meet the following criteria as applicable:

- Make sure all carpet used within the interior of the building meets the product and testing requirements of the Carpet and Rug Institute's Green Label Plus program.

- Make sure all carpet cushions used within the interior of the building meet the requirements of the Carpet and Rug Institute's Green Label program.

- Make sure all carpet adhesives meet the VOC limit of 50 g/L per IEQc4.1. The test results submitted for LEED certification shall not be more than two years old.

- Make sure all hard surface flooring (linoleum, vinyl, laminate flooring, wood flooring, ceramic flooring, rubber flooring, and wall base) are certified to meet the FloorScore™ standard by an independent third party.

- You can use FloorScore™ compliance as an **alternate credit path**: no-carpet finished flooring shall constitute 25% or more of the finished area, and 100% of the no-carpet finished area shall be FloorScore™-certified. Rooms with unfinished floors may include electrical rooms, mechanical rooms, or elevator service rooms, etc.

- Bamboo, cork, wood, and concrete floor finishes (such as stain and sealer finish) must meet the requirements of SCAQMD (South Coast Air Quality Management District) Rule 1113, Architectural Coatings, rules with an effective date of 1/1/2004.

- Grout and setting adhesive shall meet the requirements of SCAQMD (South Coast Air Quality Management District) Rule 1168 on VOC limits with an effective date of 7/1/2005, and rule amendment dated 1/7/2005.

OR

2nd Path:

Ensure all the interior flooring elements in the building meet the product and testing requirements of the California Department of Health Service Standard Practice for the Testing of Volatile Organic Emissions from Various Sources Using Small-Scale Environmental Chambers, including 2004 Addenda.

Considerations and Benefits:

See IEQc4.1

Submittals:

1) A list of interior carpet adhesives, carpet, and carpet cushions used, including VOC content of each
2) A list of tile setting adhesive, grout, finishes, and other hard surface flooring product used, including the VOC content of each tile setting grout and adhesive

Synergies:

Same list of prerequisites and credits as EAc4.1, except replace EAc4.3 with EAc4.1.

Possible Strategies and Technologies:

Make sure that you list the product certification and testing requirements in the plans or specifications. Choose products that meet the proper requirements and are tested by an independent and qualified lab or certified by the Green Label Plus program.

You can find information on the Green Label Plus program for carpets and the related VOC emission limit (microgram per square meter per hour), sample collection and testing method by Carpet and Rug Institute (CRI) with the help of California Department of Health Service (DHS) and California's Sustainable Building Task Force in Section 9, Acceptable Emission Testing for Carpet, DHS Standard Practice CA/DHS/EHLB/R-174 with a date of 7/15/2004. Refer to the following link for this document:
http://www.dhs.ca.gov

This document is also published by the Collaborative for High Performance Schools (www.chps.net) as Section 01350 Section 9 (dated 2004).

Extra Credit (Exemplary Performance):

None

Project Phase:

Design Development

LEED Submittal Phase:

Construction

Related Code or Standard:

1) Carpet and Rug Institute's Green Label Plus program
2) SCAQMD (South Coast Air Quality Management District) Rule 1168 on VOC limits with an effective date of 7/1/2005, and rule amendment dated 1/7/2005
3) SCAQMD (South Coast Air Quality Management District) Rule 1113, Architectural Coatings, rules with an effective date of 1/1/2004
4) California Department of Health Service (DHS) Standard Practice for the Testing of Volatile Organic Emissions from Various Sources Using Small-Scale Environmental Chambers, including 2004 Addenda
5) State of California Specifications, Section 01350, Standard Practice for the Testing of Volatile Organic Emissions from Various Sources Using Small-Scale Environmental Chambers, Testing Criteria
6) FloorScore™ program
7) Environmental Technology Verification (ETV) Large Chamber Test Protocol for Measuring Emissions of VOCs and Aldehydes, effective September 1999

Responsible Party:

Contractor and Architect

IEQc4.4: Low-Emitting Materials: Composite Wood and Agrifiber Products (1 Point)

Purpose:

To reduce the number of harmful, irritating, and/or odorous indoor air contaminants. Also to provide comfort and protect the well-being of the building users as well as the workers who install the materials.

Credit Path:

Use agrifiber and composite wood products that contain NO added urea-formaldehyde resins on the building interior, i.e., inside of the weatherproofing system. Use laminated adhesives that contain no added urea-formaldehyde resins to fabricate shop-applied and on-site agrifiber and composite wood products.

Composite wood products include door cores, panel substrate, MDF (medium density fiberboard), particleboard, plywood, strawboard, and wheatboard. They do not include FF&E (fit-out, furniture, and equipment) since they are not base building elements.

Agrifiber wood products include cereal straw, coconut husk, sunflower husk, sugarcane, and agricultural pruning.

Note: The above criteria do NOT apply to products covered by IEQc4.5: Low-Emitting Materials: Systems Furnishings and Seating.

Considerations and Benefits:

Refer to IEQc4.1

Submittals:

1) A list of interior agrifiber and composite wood product used, and confirmation that none of the product contain urea-formaldehyde

Synergies:

Same list of prerequisites and credits as EAc4.1, except replace EAc4.4 with EAc4.1

Possible Strategies and Technologies:

Make sure that you only list the agrifiber and composite wood products, field and shop assemblies, and laminating adhesive that contain no added urea-formaldehyde resins.

Extra Credit (Exemplary Performance):
None

Project Phase:
Design Development

LEED Submittal Phase:
Construction

Related Code or Standard:
None

Responsible Party:
Contractor and Architect

IEQc4.5: Low-Emitting Materials: System Furniture and Seating (1 Point)

Purpose:
To reduce the number of harmful, irritating, and/or odorous indoor air contaminants. Also to provide comfort and protect the well-being of building users as well as the workers who install the materials.

Credit Paths:
All system seating and furniture that has been refinished, refurbished, or manufactured within 1 year of the time of occupancy must meet one of the following criteria. Note that used and salvaged furniture older than 1 year at the time of occupancy is exempt.

1) Furniture is GREENGUARD™ Indoor Air Quality Certified℠.

OR

2) The seating products and furniture shall NOT increase indoor air concentrations by any more than the threshold limits shown in following table. The air concentrations are determined by tests conducted in an independent air quality testing laboratory per a procedure based on the EPA Environmental Technology Verification (ETV) Large Chamber Test Protocol for Measuring Emissions of VOCs and Aldehydes (September 1999) testing protocol.

Max. Indoor Air Concentrations

Chemical Contaminant	Seating	System Furniture
4-Phenylcyclohexene (4-PCH)	0.00325 mg/m^3	0.0065 mg/m^3
Formaldehyde	25 parts per billion	50 parts per billion
Total Aldehydes	50 parts per billion	100 parts per billion
Total VOCs	0. 25 mg/m^3	0. 5 mg/m^3

OR

3) The seating products and furniture shall NOT increase indoor air concentrations by any more than the threshold limits shown in table above. The air concentrations is determined by tests conducted in an independent third-party air quality testing laboratory per a procedure based on ANSI/BIFMA M7.1-2007 and ANSI/BIFMA X7.1-2007 testing protocol: The Section 5 requirements are waived for LEED projects.

Considerations and Benefits:

See IEQc4.1

Submittals:

Manufacturer's cut sheets showing seating products and furniture is

1) GREENGUARD™ Indoor Air Quality Certified℠

OR comply with

2) The EPA Environmental Technology Verification (ETV) Large Chamber Test Protocol for Measuring Emissions of VOCs and Aldehydes (September 1999)

OR

3) ANSI/BIFMA M7.1-2007 and ANSI/BIFMA X7.1-2007 testing protocol.

Synergies:

Same list of prerequisites and credits as EAc4.1, except replace EAc4.5 with EAc4.1

Possible Strategies and Technologies:

1) Use furniture that is GREENGUARD™ Indoor Air Quality Certified℠.
2) Limit **off-gassing** by complying with performance-based standards as described in Criteria 2 and 3 of **Credit Paths** above.

Extra Credit (Exemplary Performance):

None

Project Phase:

Design Development

LEED Submittal Phase:

Construction

Related Code or Standard:

1) The EPA Environmental Technology Verification (ETV) Large Chamber Test Protocol for Measuring Emissions of VOCs and Aldehydes (September 1999)
2) BIFMA International: ANSI/BIFMA M7.1-2007 and ANSI/BIFMA X7.1-2007 testing protocol
3) GREENGUARD™ Certification Program

Responsible Party:

Contractor and Architect

IEQc5: Indoor Chemical and Pollutant Source Control (1 Point)

Purpose:

To minimize building users' exposure to contaminants, pollutants, and possibly dangerous particulates.

Credit Path:

Your design should control and minimize pollutants from entering the building and prevent cross contamination between different areas:

1) Before occupancy, for mechanically ventilated buildings, install filtration media to process both outside air and return air that will be used as supply air. The filtration media shall have a MERV (Minimum Efficiency Reporting Value) value of 13 or higher. The filtration system needs to meet ANSI/ASHRAE 52.2, 1999 standards.

AND

2) In areas that may have chemical pollutants and possibly dangerous gases or particulates, such as copying/printing room, laundry/housekeeping areas, garages, prep rooms, science laboratories, art rooms, and shops of any kinds, etc., exhaust the areas sufficiently to create a negative pressure when the doors to these areas are closed. For these types of areas or spaces, provide hard lid ceilings, deck-to-deck partitions, and self-closing doors. Your exhaust rate shall be 0.50 cfm/s.f. minimum, and do not allow air re-circulation. The pressure difference between these spaces and adjacent spaces shall be at least 1 Pa (0.004 inches of water) and 5 Pa (0.02 inches of water) on average with the doors to these spaces close.

AND

3) Provide permanent entryway systems at all entries regularly used by building users and directly connected to the outdoors; these systems shall be 10 feet long minimum in the main travel direction. They are used to capture particulates and dirt, and prevent them from entering the building. You can use grilles, grates, or slotted systems that can be cleaned from below as entryway systems. Use roll-out mats only if you have contracted a service organization (or school maintenance staff for schools) to maintain them every week. CS projects with no entryway system cannot obtain this credit.

AND

4) In places where chemical concentrate and water mixing occurs (janitorial, housekeeping, science laboratories, etc.), supply containment drains for the proper disposal of hazardous liquid waste, or a closed container in a hazardous waste storage area outside the building for off-site disposal.

Considerations and Benefits:

1) **Environmental:** This credit is to prevent human contact with airborne particles and chemicals. It costs money to achieve, but is important. It can prevent particles and chemical spills from harming ecosystem and people.

2) **Financial:** Achieving this credit can increase the initial investment of a project, as well as maintenance and operation costs, but it provides valuable benefits: better health for employees, lower health care and insurance rates, reduced absenteeism, and increase productivity.

Submittals:

1) A list of entryway systems
2) A list of areas or rooms to be separated
3) A building maintenance plan including instructions for maintaining and cleaning walk-off mats and entryway systems
4) Details for hard-lid conditions and deck-to-deck partitions for rooms with contaminants
5) Visual documentation of the size and location of all walk-off mats and permanent entryway systems
6) Review negative pressure calculations at hazardous chemical areas to guarantee appropriate depressurization
7) Keep cutsheets for MERV 13 or higher filters

Synergies:

You can use filtration media to remove contaminants from the air in the construction and operation stages of projects. Refer to the following credits for high-efficiency filtration criteria:

- IEQc3.1: Construction IAQ Management Plan: During Construction
- IEQc3.2: Construction IAQ Management Plan: Before Occupancy

You can use fan systems to exhaust spaces where chemicals are used. These systems use additional energy and require commissioning. Refer to the following related prerequisites and credits:

- EAp1: Fundamental Commissioning of the Building Energy Systems
- EAp2: Minimum Energy Performance
- EAc1: Optimize Energy Performance
- EAc3: Enhanced Commissioning

Ventilation systems should be able to accommodate the filtration media. Refer to the following related prerequisite and credit:

- IEQp1: Minimum IAQ Performance
- IEQc1: Outdoor Air Delivery Monitoring

Possible Strategies and Technologies:

1) Provide a separate exhaust system for maintenance areas and facility cleaning areas to control contaminants.
2) Use permanent entryway systems to capture particulates and dirt and prevent them from entering the buildings.
3) Install high-level filtration media to process both outside air and return air that will be used as supply air. Double check your air handling units to make sure they can accommodate the required pressure drops and filters sizes.

Extra Credit (Exemplary Performance):

None

Project Phase:

Design Development

LEED Submittal Phase:

Design

Related Code or Standard:

ANSI/ASHRAE 52.2-1999 (with errata but without addenda), Methods of Testing General Ventilation Air-Cleaning Devices for Removal Efficiency by Particle Size

Responsible Party:
Contractor, MEP Engineer, and LEED AP

IEQc6.1: Controllability of Systems: Lighting (1 Point)

Purpose:
Research shows that people prefer to have control of the lighting system in spaces they occupy. The purpose of this credit is to give individuals or groups in multi-occupant spaces (such as conference areas and classrooms) a high level of controllability of the lighting system for their own preferences, well-being, and comfort.

Credit Path:

1) Give controllability of the lighting system to 90% or more of the tenant occupants for their preferences and needs.

AND

Give controllability of lighting system for ALL shared multi-occupant spaces to meet group preferences and needs.

Considerations and Benefits:

1) **Environmental:** This credit is to provide controllability of lighting systems. It costs money to achieve, but is important. It can satisfy the preferences of individuals or groups, and improve their communications, well-being, and comfort.

2) **Financial:** Achieving this credit can increase the initial investment of a project, as well as the maintenance and operation costs, but it also provides valuable benefits: lighting and HVAC savings, better communications, well-being, and comfort.

Submittals:

1) A floor plan showing the zoning, location, type of lighting controls, individual and shared work areas, and furniture layout
2) Keep design information on lighting controls, sensors, and task lighting

Synergies:
Refer to the following related prerequisites and credits:

- EAp1: Fundamental Commissioning of the Building Energy Systems
- EAp2: Minimum Energy Performance
- EAc1: Optimize Energy Performance
- EAc2: Enhanced Commissioning

- IEQc6.2: Controllability of Systems: Thermal Comfort
- IEQc8: Daylight and Views

Possible Strategies and Technologies:

1) Use task lighting and lighting controls to provide flexibility for building users.
2) Incorporate controllability into the overall lighting design.
3) Use task lighting and ambient lighting to meet the needs of building users, and still consider and control the building energy use.

Extra Credit (Exemplary Performance):

None

Project Phase:

Design Development

LEED Submittal Phase:

Design

Related Code or Standard:

None

Responsible Party:

MEP Engineer

IEQc6.2: Controllability of Systems: Thermal Comfort (1 Point)

Purpose:

Research shows that people prefer to have control of the thermal comfort system in the spaces they occupy. The purpose of this credit is to give individuals or groups in multi-occupant spaces (such as conference areas and classrooms) a high level of controllability of the thermal comfort system for their own preferences, well-being, and comfort.

Credit Path:

1) Give controllability of the thermal comfort system (for workspaces only in schools) to 50% of the tenant occupants for their preferences and needs. Use operable windows instead of comfort control if the occupants' areas are within 10 feet of either side of, and 20 feet inside of, the operable part of a window. Your building's areas of operable window shall meet ASHRAE 62.1-2007 paragraph 5.1, Natural Ventilation (with errata but without addenda), and operable window area shall be 4% or more of the total floor area. Each operable window shall have an individual control system.

AND

2) Give controllability of the thermal comfort system for ALL shared multi-occupant spaces to meet group preferences and needs.

   Include **primary factors of comfort** (air speed and air temperature, radiant temperature, and humidity) in your consideration for thermal comfort per ASHRAE Standard 55-2004. Your comfort system control shall operate at least one of the primary factors of the building user's local area to qualify for this credit.

Note:

Non-occupied areas, such as closets, equipment rooms, janitor's rooms, and storage rooms do NOT need to be included in the calculation for the controllability of thermal comfort.

Considerations and Benefits:

1) **Environmental:** The purpose of this credit is to provide controllability of thermal comfort systems. It costs money to achieve, but is important. It can satisfy the preferences of individuals or groups, and improve their communications, well-being, and comfort.

2) **Financial:** Achieving this credit can increase the initial investment of a project, as well as the maintenance and operation costs, but it also provides valuable benefits: lighting and HVAC savings, better communications, well-being, and comfort.

Submittals:

1) For controls at **individual workstations**: a list of the total quantity of thermal controls and individual workstations
2) For controls in **shared spaces**: a description of the thermal controls and a list of shared spaces

Synergies:

Refer to the following related prerequisites and credits:

- EAp1: Fundamental Commissioning of the Building Energy Systems
- EAp2: Minimum Energy Performance
- EAc1: Optimize Energy Performance
- EAc2: Enhanced Commissioning
- EAc3: Measurement and Verification

- IEQc5: Indoor Chemical and Pollutant Source Control Products
- IEQc6.1: Controllability of Systems: Lighting
- IEQc8: Daylight and Views

Possible Strategies and Technologies:

1) Incorporate controllability into the overall thermal comfort design to allow groups in shared spaces or individuals to adjust the system for their preferences and needs.
2) Refer to ASHRAE Standard 55-2004 for **primary factors of comfort** and a process for creating comfort criteria.
3) Develop control strategies to include mechanical systems alone, operable windows, or hybrid systems, which include both mechanical systems and operable windows.
4) Use local diffusers at overhead levels, desk or floor, individual thermostat controls or individual radiant panel controls.
5) Pay attention to the close relationship between acceptable **indoor air quality** (per ASHRAE Standard 62.1-2007) and **thermal comfort** (per ASHRAE Standard 55-2004).

Extra Credit (Exemplary Performance):

None

Project Phase:

Design Development

LEED Submittal Phase:
Design

Related Code or Standard:

1) ASHRAE 62.1-2007 (with errata but without addenda), Ventilation for Acceptable Indoor Air Quality
2) ASHRAE Standard 55-2004, Thermal Environmental Conditions for Human occupancy

Responsible Party:
MEP Engineer

IEQc7.1: Thermal Comfort: Design (1 Point)

Purpose:
To protect the well-being of building users and to increase their productivity via a comfortable thermal environment.

Credit Path for NC, Schools, and CS:
Design HVAC systems per ASHRAE Standard 55-2004, Thermal Comfort Conditions for Human Occupancy (with errata but without addenda). You need to show design compliance per Section 6.1.1 Documentation. Use the Predicted Mean Vote (PMV) model.

Considerations and Benefits:

1) **Environmental:** The purpose of this credit is to provide thermal comfort via design. It costs money to achieve, but is important. It can improve occupants' well-being and comfort.

2) **Financial:** Achieving this credit can increase the initial investment of a project, as well as the maintenance and operation costs, but also provides valuable benefits: lighting and HVAC savings, as well as occupant well-being and comfort.

Submittals:

1) Documentation of OPR, including assumptions for occupant clothing and activity level, and comfort criteria
2) Building systems operational procedure and summary, including general information, changeover schedules, recommended seasonal set points, maintenance and operation instructions, inspection and maintenance schedules, building controls, other environmental control systems, etc.
3) Documentation of mechanical engineer's BOD, including HVAC load calculations, diversity consideration, and design assumption
4) Keep design plans, lists, and other documentation of all terminal units and registers, including radiant value, flow, type, and any factors that have significant impact on thermal comfort, all occupant-adjustable control locations, and notation of spaces outside comfort-controlled areas.

Synergies:

Thermal comfort is affected by personal factors (clothing and metabolic rate), environment (air speed, air temperature, radiant temperature, and relative humidity). Thermal comfort can be controlled through passive (natural ventilation) and/or active (HVAC) systems. A combination of both systems can achieve optimum comfort levels and save energy.

Refer to the following related prerequisite and credits:

- EAp2: Minimum Energy Performance
- EAc1: Optimize Energy Performance
- EAc3: Measurement and Verification

For commissioning of thermal comfort features, refer to the following related prerequisite and credit:

- EAp1: Fundamental Commissioning of the Building Energy Systems
- EAc2: Enhanced Commissioning

The following prerequisite and credit also relate to occupants' comfort:

- IEQp1: Minimum IAQ Performance
- IEQc2: Increased Ventilation

Possible Strategies and Technologies:

Set up comfort criteria per ASHRAE Standard 55-2004. Evaluate the primary factors of comfort (air speed and air temperature, radiant temperature and humidity) and relate them to EQp1, IEQc1, and IEQc2.

Extra Credit (Exemplary Performance):

None

Project Phase:

Design Development

LEED Submittal Phase:

Design

Related Code or Standard:

ANSI/ASHRAE Standard 55-2004

Responsible Party:

MEP Engineer

IEQc7.2: Thermal Comfort: Verification (1 Point)

Purpose:

To follow up and evaluate the building's thermal comfort.

Credit Path:

Conduct a thermal comfort survey of tenant occupants within 6 to 18 months after occupancy. The survey should be anonymous, and it should identify the problems related to thermal comfort and overall satisfaction with the thermal system performance. If fewer than 80% of the building occupants are satisfied, then develop a plan to correct the problems. Corrective actions should include measuring the environmental variables in the problem areas per ASHRAE Standard 55-2004 (with errata but without addenda).

Install a permanent monitoring system to make sure the building performs as designed and meets the comfort criteria per IEQc7.1.

Residential projects do not qualify for this credit.

Considerations and Benefits:

1) **Environmental:** This credit is to verify thermal comfort. It costs money to achieve, but is important. It can improve occupants' well-being and comfort.

2) **Financial:** Achieving this credit can increase the maintenance and operation costs, but it also provides valuable benefits: lighting and HVAC savings, well-being, and comfort.

Submittals:

1) Create and administer a thermal comfort survey for building occupants
2) If 20% or more of the building occupants are not satisfied with the thermal comfort, generate a plan for corrective action

Synergies:

Thermal comfort is affected by personal factors (clothing and metabolic rate), environment (air speed, air temperature, radiant temperature, and relative humidity). Thermal comfort can be controlled through passive (natural ventilation) and/or active (HVAC) systems. A combination of both systems can achieve optimum comfort levels and save energy.

Refer to the following related prerequisite and credits:

- EAp1: Fundamental Commissioning of the Building Energy Systems
- EAp2: Minimum Energy Performance
- EAc1: Optimize Energy Performance
- EAc2: Enhanced Commissioning
- EAc3: Measurement and Verification

The following prerequisite and credits also relate to occupants' comfort:

- IEQp1: Minimum IAQ Performance
- IEQc2: Increased Ventilation
- IEQc6.2: Controllability of Systems: Thermal Comfort
- IEQc7.1: Thermal Comfort: Design

Possible Strategies and Technologies:

Use the principles in ASHRAE Standard 55-2004 to develop criteria for thermal comfort, and a plan for monitoring and correcting performance of the thermal comfort systems.

Extra Credit (Exemplary Performance):

None

Project Phase:
Design Development

LEED Submittal Phase:
Design

Related Code or Standard:
ANSI/ASHRAE Standard 55-2004

Responsible Party:
MEP Engineer

IEQc8.1: Daylight and Views: Daylight (1-2 Point)

Purpose:
To connect the inside and outside of the building by introducing views and daylight to the regularly occupied areas inside the tenant space.

Credit Paths:
Minimum criteria to obtain the points:

% of Daylight in Spaces	Points
75%	1
90%	2

1) **Prescriptive**

For side-lighting daylight area:

$$0.15 < VLT \times WFR < 0.18$$

VLT is visual light transmittance
WFR is window-to-floor area ratio

- The window area must be a minimum of 30 inches above the floor to be included in the calculation.
- The horizontal distance from the edge of the daylight area to the bottom of the wall (where the window is located) shall be less than two times the vertical height from the bottom of the wall to the top of window mullion.
- If you draw a line from the edge of the side-lighting daylight area to the top of window glass, the ceiling shall be above this line so as not to block daylight.

For top-lighting daylight area:

- The top-lighting daylight area under the skylight is defined as the area directly under the outline of the skylight PLUS the lesser of:

70% of the ceiling height

OR
50% of the distance to the closest of the adjacent skylights

OR
The distance to any permanent opaque partition (show VLT for transparent partitions) that is at a distance of 70% or more of the gap between the ceiling and the top of the partition.

- Skylight coverage shall be 3% to 6% of the roof area with a VLT of at least 0.5
- The distance between skylights shall be 1.4 times the ceiling height or less.
- If you use a skylight diffuser, it shall have a measured haze value of 90% or more when tested per ASTM D1003. In your design, avoid direct sightlines to the skylight diffuser.

OR

2) Computer Simulation

Use a computer simulation to show that a minimum of 75% (1 point) or 90% (2 points) of the regularly occupied areas have at least 25 horizontal footcandles (fc) and do not exceed 500 fc of daylight illumination. This simulation should be based on a clear sky situation, on September 21, at 9 a.m. and 3 p.m., at 30 inches above the finished floor. If your design incorporates view-preserving automated shades for glare control, you just need to meet the 25 fc minimum requirement.

OR

3) Measurement

Use records of measurements of indoor lights to show that a minimum of 75% (1 point) or 90% (2 points) of the regularly occupied areas have at least 25 horizontal footcandles of daylight illumination. Take the measurements on a 10-foot grid and mark the measurements on the floor plans.

For all three options, only apply the s.f. of the portions of the spaces with at least 25 horizontal footcandles of daylight illumination towards the 75% required areas.

For all three options, use control devices or redirect daylight to avoid high-contrast conditions and prevent impeding visual tasks. Exceptions to this rule will be considered based on their merits.

4) Combination

Any combination of the three paths above can be used. If the combination results in 75% of the regularly occupied areas meeting the requirements, your project can get 1 point. If the combination results in 90% of the regularly occupied areas meeting the requirements, your project can get 2 points.

In all options, only the area related to the spaces or rooms meeting the requirements can be included in the calculation.

In all options, glare control devices can be used to maximize daylight for interior spaces while avoiding high-contrast conditions.

Note:

1) Include all regularly occupied spaces like cafeterias, meeting rooms, and office spaces, but do NOT include spaces such as closets, utility rooms, bathrooms, copy rooms, laundry rooms, restrooms, equipment rooms, janitor's rooms, or storage rooms in the area calculations for daylight and views.
2) Include window areas between 2'-6" AFF and 7'-6" AFF, and glazing areas above 7'-6" AFF. Do NOT include areas for any openings below 2'-6" AFF.

Considerations and Benefits:

1) **Environmental:** The purpose of this credit is to introduce views and daylight to the regularly occupied areas. It can improve occupants' well-being and comfort as well as save 50% to 80% of lighting energy costs. Building and room orientation, balance of heat loss and gain, glare control, light shelves, shading devices, photo responsive controls, atrium, courtyards, sizes and spacing of windows, colors and types of glazing, locations of interior walls, and reflectance of interior finishes can all contribute to this credit.

   Large areas of untreated glazing can cause illusions and birds to fly into the glazing. Up to 1 billion birds may die this way per year. Visual markers, fritt or etched patterns, exterior shading device, as well as changes in materials, colors, textures, and opacity can reduce the chances of bird collisions.

2) **Financial:** Large areas of windows can increase heat loss or gain, special glazing may cost more. It can increase the maintenance and operation costs to achieve this credit, but it can also provide valuable benefits: lighting and HVAC savings, improved occupants' satisfaction, well-being, and comfort. Employee salaries are many times greater than the extra costs to achieve this credit.

Submittals:

1) Create floor plans, elevations, sections, etc. to show the glare control methods used
2) Keep floor plans, elevations, sections, etc. to show a qualifying amount of daylight for regularly occupied spaces
3) A spreadsheet to document daylight factors outlined in calculations, and changes in design
4) Update the computer model continuously if using the simulation method

Synergies:

Increasing window size to gain more daylight can also increase views. Refer to the following related credit:

- IEQc8.2: Daylight and Views

Larger windows have a direct impact on building energy performance. Daylighting controls can maximize energy savings. Refer to the following related prerequisite and credits:

- EAp2: Minimum Energy Performance
- EAc1: Optimize Energy Performance
- IEQc6: Controllability of Systems

Possible Strategies and Technologies:

1) Use interior and exterior permanent shading devices, adjust building orientation, increase building perimeter, design shallow floor plates, properly design the size, number and locations of building openings, consider neighboring buildings and trees, use courtyards, atriums, clerestory

windows, north-facing skylight, interior light (reflecting) shelf, and use automatic photocell-based control and high performance glazing to maximize daylight for interior spaces while avoiding high-contrast conditions.

2) Use light fixtures with sensors and dimming controls.
3) Use a computer model, physical model, or manual calculation to evaluate daylight footcandle levels.

Extra Credit (Exemplary Performance):
None

Project Phase:
Schematic Design

LEED Submittal Phase:
Design

Related Code or Standard:
ASTM D1003-07E1, Standard Testing Method for Haze and Luminous Transmittance of Transparent Plastics

Responsible Party:
Lighting Designer, Architect, and LEED AP

IEQc8.2: Daylight and Views: Views (1 Point)

Purpose:
To connect the inside and outside of the building by introducing views and daylight to regularly occupied areas inside the building.

Credit Path for NC, Schools, and CS:
Provide views (direct line of sight to the outside) for 90% of the regularly occupied spaces. The line of sight should be via vision glazing at a height between 2'-6" and 7'-6" AFF. An area needs to meet the following two criteria to be included as an area with a view:

1) In plan view, be able to draw a line of sight from the area to the perimeter vision glazing (between 2'-6" and 7'-6" AFF).
2) In section view, draw a line of sight from the area to the perimeter vision glazing. Use 42" as the average seated eye height.

You can draw a line of sight through interior glazing. If at least 75% of a private office has a view, then you can count the entire area of the office as an area with a view. If less than 75% of a private office has a view, then you can only count the actual area. If a space has multiple occupants, then you can only count the actual areas with a view.

Considerations and Benefits:
Refer to IEQc8.1

Submittals:

1) Keep floor plans, elevations, sections, etc. to show the locations of regularly occupied spaces with views
2) A spreadsheet to document the view area outlined in calculations, and any changes in design

Synergies:

Same as IEQc8.1, except IEQc8.1 is replaced with IEQc8.2.

Possible Strategies and Technologies:

1) Design the space to maximize views and daylight.
2) Use interior glazing, interior shading devices, low partition heights, and automatic photo-cell controls.

Extra Credit (Exemplary Performance):

If two out of the following four items are achieved, you can obtain an extra ID point:

1) At least 90% of the regularly occupied spaces have access to unobstructed views within a distance of three times the vision window head height.
2) At least 90% of the regularly occupied spaces have more than one sight line to vision glazing in different directions at 90 degrees or more apart.
3) At least 90% of the regularly occupied spaces have views to two of the three choices: human activities; vegetation; objects 70 feet or more from the exterior of the glazing.
4) At least 90% of the regularly occupied spaces have access to views with a view factor of 3 or more. Refer to: http://h-m-g.com/downloads/Daylighting/day_registration_form.htm

Project Phase:

Schematic Design

LEED Submittal Phase:

Design

Related Code or Standard:

None

Responsible Party:

Lighting Designer, Architect, and LEED AP

IEQ Summary and Mnemonics:

IEQ Credit Name	Extra Credit	Responsible Party
IEQp1: Minimum IAQ Performance	0	MEP Engineer
IEQp2: Environmental Tobacco Smoke (ETS) Control	0	LEED AP, Owner, and MEP Engineer
IEQc1: Outdoor Air Delivery Monitoring	0	MEP Engineer
IEQc2: Increased Ventilation	0	Contractor and MEP Engineer

IEQc3.1: Construction IAQ Management Plan: During Construction	0	Contractor and MEP Engineer
IEQc3.2: Construction IAQ Management Plan: Before Occupancy	0	Contractor and MEP Engineer
IEQc4.1: Low-Emitting Materials: Adhesive and Sealants	0	Contractor and Architect
IEQc4.2: Low-Emitting Materials: Paints and Coatings	0	Contractor and Architect
IEQc4.3: Low-Emitting Materials: Carpet Systems	0	Contractor and Architect
IEQc4.4: Low-Emitting Materials: Composite Wood and Agrifiber Products	0	Contractor and Architect
IEQc4.5: Low-Emitting Materials: System Furniture and Seating	0	Contractor and Architect
IEQc5: Indoor Chemical and Pollutant Source Control	0	Contractor, MEP Engineer, and LEED AP
IEQc6.1: Controllability of Systems: Lighting	0	MEP Engineer
IEQc6.2: Controllability of Systems: Thermal Comfort	0	MEP Engineer
IEQc7.1: Thermal Comfort: Design	0	MEP Engineer
IEQc7.2: Thermal Comfort: Verification	0	MEP Engineer
IEQc8.1: Daylight and Views: Daylight	0	Lighting Designer, Architect, and LEED AP
IEQc8.2: Daylight and Views: Views	If you achieve two out of the four items below, you can obtain an extra ID point: 1) At least 90% of the regularly occupied spaces have access to unobstructed views within a distance of three times the vision window head height. 2) At least 90% of the regularly occupied spaces have more than one sight line to vision glazing in different directions at 90 degrees or more apart. 3) At least 90% of the regularly occupied spaces have views to two of the three choices: human activities; vegetation; objects 70 feet or more from the exterior of the glazing. 4) At least 90% of the regularly occupied spaces have access to views with a view factor of 3 or more.	Lighting Designer, Architect, and LEED AP

Mnemonics:
ME
Oh! Ian Catches LIST D (See bold and underlined letters in credit names on table above also). Of course, you can create your own **mnemonics** to help you memorize them.

F. Innovation and Design Process (ID)

Because this credit category is relatively simple, we omit the overall discussion.

IDc1: Innovation in Design (1-5 points)

Purpose:
To reward design teams and projects for innovative performance in green building categories NOT covered by LEED and/or exceptional performance above and beyond the LEED requirements.

Credit Paths:
You can obtain credits through any combinations of the following paths:

1) **Innovation in Design (1-5 points)**
   In writing, provide the intent, requirements, proposed submittals, and design approach for the proposed innovation credit.
   1) You need to show innovative performance in green building categories NOT covered by an existing LEED credit.
   2) The GBCI will award one point for each innovation achieved.
   3) The **criteria** are: quantitative performance improvements for environmental benefit; comprehensive specification or process (applicable to the entire project, not just a portion); substantially better than typical sustainable practice and applicable to other projects.

OR

2) **Exemplary Performance (1-3 points)**
   1) You need to show exceptional performance above and beyond the LEED requirements for an existing credit. For example, for a number of LEED credit categories, you can get one innovation point by doubling the requirements or reaching the next threshold.
   2) The GBCI will award one point for each exemplary performance achieved.
   3) A maximum of three points can be achieved through this path.

   Note: The benefits from different paths must be additive.

Considerations and Benefits:
Sustainable design originates from creative thinking and innovative strategies. LEED points to reward this kind of thinking and strategies will benefit the environment and all of us.

Submittals:

1) Documentation for the process of the project team to achieve environmental benefits or exceptional performance above and beyond the LEED requirements in an existing credit, or innovative performance in green building categories NOT covered by an existing LEED credit
2) Track implementation and development of the innovative and exceptional strategies used

Synergies:

Refer to the extra credit (exemplary performance) section of each credit.

Possible Strategies and Technologies:

1) Try to substantially exceed the LEED performance credit (such as WE or IEQ).
2) Use measures that can show a quantifiable health and/or environmental benefits, and a comprehensive approach.
3) Try to select innovative performance that is applicable for other projects.
4) If an innovation point is awarded, the GBCI does not guarantee it will be awarded for future projects.
5) You need to write a separate submittal and narrative statement for each point.

Extra Credit (Exemplary Performance):

None (IDc1 IS the extra ID credit for other LEED credit categories).

Project Phase:

Varies

LEED Submittal Phase:

Construction or Design

Related Code or Standard:

None

Responsible Party:

Varies (All disciplines, Owner, and Contractor)

IDc2: LEED Accredited Professional (1 point)

Purpose:

To encourage and support the design integration dictated by LEED and to streamline the certification and application process.

Credit Path:

To get this credit, at least one principal project team member (not a junior team member) shall be a LEED AP (Accredited Professional).

Considerations and Benefits:
LEED AP can lead the project's LEED application process and educate other team members about LEED.

Submittals:
Confirmation of the LEED AP status of at least one principal of the team.

Synergies:
None

Possible Strategies and Technologies:
1) Try to assign the LEED AP within your organization as the facilitator for the integrated design and construction process.
2) Hire a LEED AP with LEED CI experience
3) Educate your team early about the LEED rating system and green buildings.

Extra Credit (Exemplary Performance):
None

Project Phase:
ALL phases

LEED Submittal Phase:
Construction

Related Code or Standard:
LEED AP. Refer to www.GBCI.org

Responsible Party:
LEED AP, Owner, and Contractor

ID Summary and Mnemonics:

ID Credit Name	Extra Credit	Responsible Party
IDc1: Innovation in Design (1-5 points)	None (ID credits ARE the extra ID credits for other LEED credit categories).	Varies (All disciplines, Owner, and Contractor)
IDc2: LEED Accredited Professional (1 point)	Same as above	LEED AP, Owner, and Contractor

G. Regional Priority (RP)

Because this credit category is relatively simple, we omit the overall discussion.

RPc1: Regional Priority (1-4 points)

Purpose:
To encourage the achievement of credit for addressing regional environmental priorities.

Credit Paths:
Refer to the RP credit database at
http://www.usgbc.org

You can obtain 1 to 4 credits out of the 6 potential RP points. You can choose which 4 credits you want to pursue for your project.

Projects outside of the US cannot obtain RP credits at this time.

Considerations and Benefits:
Refer to each specific credit gaining RP point(s).

Submittals:
Refer to the RP database at www.usgbc.org and LEED Online

Synergies:
Refer to the RP database at www.usgbc.org

Possible Strategies and Technologies:
1) The available RP credits will show up at LEED Online after you enter your project's specific information like zip code.
2) Coordinate with other credits and choose which 4 RP credits you want to pursue for your project.

Extra Credit (Exemplary Performance):
None

Project Phase:
ALL phases

LEED Submittal Phase:
Design

Related Code or Standard:
None

Responsible Party:
LEED AP

Chapter 5

Summary & Review Tables

1. Reference Codes, Standards and related LEED credits

Advanced Buildings™ Core Performance™ Guide by New Buildings Institute:
EAc1.3: Optimize Energy Performance: HVAC (5-10 points)

ASHRAE 52.2-1999 (with errata but without addenda):
IEQc3.1: Construction IAQ Management Plan: During Construction (1 point)
IEQc5: Indoor Chemical and Pollutant Source Control (1 point)

ASHRAE Standard 55-2004:
IEQc6.2: Controllability of Systems: Thermal Comfort
IEQc7.1: Thermal Comfort: Design (1 point)
IEQc7.2: Thermal Comfort: Verification (1 point)

ASHRAE 62.1-2007:
IEQp1: Minimum IAQ Performance (Mandatory, 0 points)
IEQc1: Outdoor Air Delivery Monitoring (1 point)
IEQc2: Increased Ventilation (1 point)
IEQc6.2: Controllability of Systems: Thermal Comfort (1 point)

ASHRAE/IESNA Standard 90.1-2007 (with errata but without addenda):
SSc1, Ch2F11: On-Site Renewable Energy (1 to 2 point)

EAp2: Minimum Energy Performance (Mandatory, 0 points)
EAc1.1: Optimize Energy Performance: Lighting Power (1-5 points)
EAc1.3: Optimize Energy Performance: HVAC (5-10 points)
EAc1.4: Optimize Energy Performance: Equipment and Appliances (1-4 points)

American National Standards Institute (ANSI):
SSc1, Ch2F11: On-Site Renewable Energy (1 to 2 point)

EAp2: Minimum Energy Performance (Mandatory, 0 points)
EAc1.1: Optimize Energy Performance: Lighting Power (1-5 points)
EAc1.3: Optimize Energy Performance: HVAC (5-10 points)
EAc1.4: Optimize Energy Performance: Equipment and Appliances (1-4 points)

IEQp1: Minimum IAQ Performance (Mandatory, 0 points)
IEQp2: Environmental Tobacco Smoke (ETS) Control (Mandatory, 0 points)
IEQc1: Outdoor Air Delivery Monitoring (1 point)

IEQc4.5: Low-Emitting Materials: System Furniture and Seating (1 point)
IEQc5: Indoor Chemical and Pollutant Source Control (1 point)
IEQc7.1: Thermal Comfort: Design (1 point)
IEQc7.2: Thermal Comfort: Verification (1 point)

American Society for Testing and Materials (ASTM):
SSc1, Ch2F1: Brownfield Redevelopment (1 point)
SSc1, Ch2F4: Heat Island Effect: Non-Roof (1 point)
IEQp2: Environmental Tobacco Smoke (ETS) Control (Mandatory, 0 points)
IEQc8.1: Daylight and Views: Daylight (1-2 points)

Best Management Practice (BMP):
SSc1, Ch2F3: Storm Water Design: Quality Control (1 point)

BIFMA International: ANSI/BIFMA M7.1-2007 and ANSI/BIFMA X7.1-2007 Testing Protocol:
IEQc4.5: Low-Emitting Materials: System Furniture and Seating (1 point)

California Department of Health Service (DHS):
IEQc4.1: Low-Emitting Materials: Adhesive and Sealants (1 point)
IEQc4.2: Low-Emitting Materials: Paints and Coatings (1 point)
IEQc4.3: Low-Emitting Materials: Carpet Systems (1 point)

California Energy Code Title 24 (T-24):
EAp2: Minimum Energy Performance (Mandatory, 0 points)
EAc1.1: Optimize Energy Performance: Lighting Power (1-5 points)

California Integrated Waste Management Board 1999 (CIWMB 1999):
MRp1: Storage and Collection of Recyclables (Mandatory, 0 points)

Carbon Trust "Good Practice Guide 237":
IEQc2: Increased Ventilation (1 point)

Carpet and Rug Institute's (CRI) Green Label Plus Program:
IEQc4.3: Low-Emitting Materials: Carpet Systems (1 point)

Center for Resource Solution (CRS) Green-e Program:
SSc1, Ch2F11: On-Site Renewable Energy (1 to 2 points)
EAc4: Green Power (5 points)

Chartered Institution of Building Service Engineers (CIBSE):
IEQc2: Increased Ventilation (1 point)

Commercial Building Energy Consumption Survey (CBECS):
SSc1, Ch2F11: On-Site Renewable Energy (1 to 2 points)
EAc4: Green Power (5 points)

Comprehensive Environmental Response, Compensation and Liability Act (CERLA):
SSc1, Ch2F1: Brownfield Redevelopment (1 point)

Efficiency Valuation Organization (EVO):
EAc3: Measurement and Verification (2-5 points)

Energy Information Administration (EIA):
SSc1, Ch2F11: On-Site Renewable Energy (1 to 2 points)

Energy Policy Act (EPAct) of 1992 (and as amended) and Energy Policy Act (EPAct) of 2005:
SSc1, Ch2F9: Innovative Wastewater Technologies (Indoor Water, 2 points)
SSc1, Ch2F10: Water Use Reduction: 30% Reduction (1 point)
WEp1: Water Use Reduction (Indoor Water, Mandatory, 0 points)
WEc1: Water Use Reduction (Indoor Water, 6 to 11 points)

Energy Star Program, Target Finder Rating Tool:
EAp2: Minimum Energy Performance (Mandatory, 0 points)

Environmental Protection Agency (EPA):
SSc1, Ch2F1: Brownfield Redevelopment (1 point)
SSc1, Ch2F3: Storm Water Design: Quality Control (1 point)

WEp1: Water Use Reduction (Indoor Water, Mandatory, 0 points)

EAp3: Fundamental Refrigerant Management

MRc2: Construction Waste Management (1-2 points)

IEQc3.2: Construction IAQ Management Plan: Before Occupancy (1 point)
IEQc4.5: Low-Emitting Materials: System Furniture and Seating (1 point)

FloorScore™ program:
IEQc4.3: Low-Emitting Materials: Carpet Systems (1 point)

Forest Stewardship Council (FSC):
MRc7: Certified Wood (1 point)

Green Seal Standard:
IEQc4.1: Low-Emitting Materials: Adhesive and Sealants (1 point)
IEQc4.2: Low-Emitting Materials: Paints and Coatings (1 point)

GREENGUARD™ Certification Program:
IEQc4.5: Low-Emitting Materials: System Furniture and Seating (1 point)

International Associations of Plumbing and Mechanical Officials, Publication IAPMO/American National Standards Institute UPC 1-2006, Uniform Plumbing Code 2006, Section 402.0, Water-Conserving Fixtures and Fittings:
SSc1, Ch2F9: Innovative Wastewater Technologies (Indoor Water, 2 points)

International Code Council (ICC), International Plumbing Code (IPC) 2006, Section 604, Design of Building Water Distribution System:

WEp1: Water Use Reduction (Indoor Water, Required for NC, CS, and Schools)
WEp1: Water Use Reduction (Indoor Water, Mandatory, 0 points)

International Organization for Standardization (ISO):

MRc4: Recycled Content: post-consumer + ½ pre-consumer (1-2 points)

International Performance Measure and Verification Protocol (IPMVP), Volume III, Efficiency Valuation Organization (EVO) 30000.1-2006, Concepts and Options for Determining Energy Savings in New Construction:

EAc3: Measurement and Verification (2-5 points)

Montreal Protocol 1987:

EAp3: Fundamental Refrigerant Management (Mandatory, 0 points)

Sheet Metal and Air Conditioning National Contractors Association (SMACNA):

IEQc3.1: Construction IAQ Management Plan: During Construction (1 point)

South Coast Air Quality Management District (SCAQMD):

IEQc4.1: Low-Emitting Materials: Adhesive and Sealants (1 point)
IEQc4.2: Low-Emitting Materials: Paints and Coatings (1 point)
IEQc4.3: Low-Emitting Materials: Carpet Systems (1 point)

State of California Specifications, Section 01350, Standard Practice for the Testing of Volatile Organic Emissions from Various Sources Using Small-Scale Environmental Chambers, Testing Criteria:

IEQc4.3: Low-Emitting Materials: Carpet Systems (1 point)

U.S. Department of Energy:

SSc1, Ch2F11: On-Site Renewable Energy (1 to 2 point)
EAp2: Minimum Energy Performance (Mandatory, 0 points)
EAc4: Green Power (5 points)

2. Project Phase and Related LEED Prerequisites and Credits

The information in this section will not only help you to pass the LEED exam, but also help you to do real LEED projects. For example, you can use the information in this chapter to look up **which prerequisites or credits need to be addressed in each phase of your project**. This can be extremely helpful, simplify your tasks, and make you life much easier.

Pre-Design (PD)

SSc1, Ch1: Choose a building that is LEED Certified (5 points)
SSc1, Ch2F1: Brownfield Redevelopment (1 point)
SSc1, Ch2F2: Storm Water Design: Quantity Control (1 point)
SSc1, Ch2F3: Storm Water Design: Quality Control (1 point)

SSc1, Ch2F4: Heat Island Effect: Non-Roof (1 point)
SSc1, Ch2F5: Heat Island Effect: Roof (1 point)
SSc1, Ch2F7: Water Efficient Landscaping: Reduced by 50% (Outdoor Water, 2 points)
SSc1, Ch2F8: Water Efficient Landscaping- No Potable Water Use or No Irrigation (Outdoor Water, 2 points in addition to Feature 7)
SSc1, Ch2F9: Innovative Wastewater Technologies (Indoor Water, 2 points)
SSc1, Ch2F10: Water Use Reduction: 30% Reduction (1 point)
SSc1, Ch2F11: On-Site Renewable Energy (1 to 2 points)
SSc1, Ch2F12: Other Quantifiable Environmental Performance (1 point)

SSc2: Development Density and Community Connectivity (6 points)

SSc3.1: Alternative Transportation: Public Transportation Access (6 points)

Mnemonics: For SS categories, all credits and features occur at the Pre-Design Stage, except SSc1, Ch2F6, SSc3.2 and SSc3.3, which all happen at the Schematic Design Stage.

WEp1: Water Use Reduction (Indoor Water, Mandatory, 0 points)

MRc1.1: Tenant Space: Long-Term Commitment (1 point)
MRc1.2: Building Reuse: Maintain Interior Nonstructural Components (1-2 points)

IEQp2: Environmental Tobacco Smoke (ETS) Control (Mandatory, 0 points)

Schematic Design (SD)

SSc1, Ch2F6: Light Pollution Reduction (1 point)
SSc3.2: Alternative Transportation: Bicycle Storage and Changing Rooms (2 points)
SSc3.3: Alternative Transportation: Parking Availability (2 points)

WEc1: Water Use Reduction (Indoor Water, 6 to 11 points)

EAp2: Minimum Energy Performance (Mandatory, 0 points)
EAc1.1: Optimize Energy Performance: Lighting Power (1-5 points)
EAc1.2: Optimize Energy Performance: Lighting Controls (1-3 points)
EAc1.3: Optimize Energy Performance: HVAC (5-10 points)

MRp1: Storage and Collection of Recyclables (Mandatory, 0 points)
MRc4: Recycled Content: post-consumer + ½ pre-consumer (1-2 points)
MRc5: Regional Materials (1-2 points)
MRc6: Rapidly Renewable Materials (1 point)
MRc7: Certified Wood (1 point)

Mnemonics: For MR category, MRp1, MRp4, MRp5, MRp6, and MRp7 (see above) occur at the Schematic Design Stage, MRc1.1 and MRc1.2 happen at the Pre-Design Stage, MR c2 and MRc3.1 happen at the Construction Documents Stage, and MRC3.2 happens at Design Development Stage.

IEQp1: Minimum IAQ Performance (Mandatory, 0 points)
IEQc1: Outdoor Air Delivery Monitoring (1 point)
IEQc2: Increased Ventilation (1 point)
IEQc8.1: Daylight and Views: Daylight (1-2 point)
IEQc8.2: Daylight and Views: Views (1 point)

Mnemonics: For IEQ category, IEQp1, IEQc1, IEQc2, IEQc8.1, and IEQc8.2 occur at the Schematic Design Stage, IEQc3.1, and IEQc3.2 occur at the Construction Administration Stage, and all others happen at the Design Development Stage.

Design Development (DD)

EAp3: Fundamental Refrigerant Management (Mandatory, 0 points)
EAc3: Measurement and Verification (2-5 points)

MRc3.2: Materials Reuse: Furniture and Furnishings (1 point)

IEQc4.1: Low-Emitting Materials: Adhesive and Sealants (1 point)
IEQc4.2: Low-Emitting Materials: Paints and Coatings (1 point)
IEQc4.3: Low-Emitting Materials: Carpet Systems (1 point)
IEQc4.4: Low-Emitting Materials: Composite Wood and Agrifiber Products (1 point)
IEQc4.5: Low-Emitting Materials: System Furniture and Seating (1 point)
IEQc5: Indoor Chemical and Pollutant Source Control (1 point)
IEQc6.1: Controllability of Systems: Lighting (1 point)
IEQc6.2: Controllability of Systems: Thermal Comfort (1 point)
IEQc7.1: Thermal Comfort: Design (1 point)
IEQc7.2: Thermal Comfort: Verification (1 point)

Construction Documents (CD)

MRc2: Construction Waste Management (1-2 points)
MRc3.1: Materials Reuse (1-2 points)

Construction Administration (CA)

EAp1: Fundamental Commissioning of the Building Energy Systems (Mandatory, 0 points)
EAc2: Enhanced Commissioning (5 points)

IEQc3.1: Construction IAQ Management Plan: During Construction (1 point)
IEQc3.2: Construction IAQ Management Plan: Before Occupancy (1 point)

Occupation/Operation (OO) or Operation and Maintenance (OM)

EAc1.4: Optimize Energy Performance: Equipment and Appliances (1-4 points)
EAc4: Green Power (5 points)

All or Various Stages:

IDc1: Innovation in Design (1-5 points)
IDc2: LEED Accredited Professional (1 point)

RPc1: Regional Priority (1-4 points)

Mnemonics: For ID & RP category, all credits can occur at various stages.

3. LEED Submittal Phase and Related LEED Prerequisites and Credits

The following LEED credits can ONLY be part of Construction Phase Submittal for LEED Certification; they can NOT be submitted as part of Design Phase Submittal:

EAp1: Fundamental Commissioning of the Building Energy Systems (Mandatory, 0 points)
EAc2: Enhanced Commissioning (5 points)
EAc4: Green Power (5 points)

MRc1.2: Building Reuse: Maintain Interior Nonstructural Components (1-2 points)
MRc2: Construction Waste Management (1-2 points)
MRc3.1: Materials Reuse (1-2 points)
MRc3.2: Materials Reuse: Furniture and Furnishings (1 point)
MRc4: Recycled Content: post-consumer + ½ pre-consumer
MRc5: Regional Materials (1-2 points)
MRc6: Rapidly Renewable Materials (1 point)
MRc7: Certified Wood (1 point)

Mnemonics: ALL MR prerequisites and credits, except for MRp1 & MR c1.1, are Construction Submittals for LEED Certification

IEQc3.1: Construction IAQ Management Plan: During Construction (1 point)
IEQc3.2: Construction IAQ Management Plan: Before Occupancy (1 point)
IEQc4.1: Low-Emitting Materials: Adhesive and Sealants (1 point)
IEQc4.2: Low-Emitting Materials: Paints and Coatings (1 point)
IEQc4.3: Low-Emitting Materials: Carpet Systems (1 point)
IEQc4.4: Low-Emitting Materials: Composite Wood and Agrifiber Products (1 point)
IEQc4.5: Low-Emitting Materials: System Furniture and Seating (1 point)

IDc1: Innovation in Design (1-5 points)
Note: This credit is a special case; it can be submitted in the Construction Phase OR the Design Phase, depending on each credit's requirements.)

IDc2: LEED Accredited Professional (1 point)

Chapter 6

LEED AP ID+C Exam Sample Questions, Mock Exam, Answers and Exam Registration (Including Both Section One and Section Two)

I. LEED AP ID+C Exam sample questions

Use these sample questions to prepare for the mock exam and the real exam. They will give you an idea what the USGBC is looking for on the LEED AP ID+C Exam, and how the questions will be asked. These sample questions are quite easy. If you can answer 85% of the sample questions correctly, you are ready to take the mock exam in this chapter. The 85% passing score is based on feedback from previous readers. You need to read the study material in the previous chapters one or more times to become familiar with it and then take the mock exam. You really need to read this book several times and MEMORIZE the important information before you do the sample questions and mock exam. Just like on the real exam, sometimes a question may ask you to pick two or three correct answers out of four, or four correct answers out of five (some LEED exam questions have five choices). This means that if you do not know any one of the correct answers, you will probably get the overall answer wrong. You need to know the LEED system very well to get the correct answer.

Exam Part One sample questions:

1. With regard to the Optimize Energy Performance credit, which of the following statements is correct?
   a. Compare your building performance to the baseline building performance.
   b. Compare baseline building performance to ASHRAE Standard 90.1-2007 (with errata but without addenda).
   c. Compare baseline building performance to ASHRAE Standard 90.1-2004.
   d. Compare baseline building performance to ASHRAE Standard 90.1-2003.

2. Which of the following factors does not improve human comfort?
   a. Air temperature.
   b. Ventilation.
   c. Radiation exchange.
   d. All of the above.

3. The standard used for Measurement and Verification is:
   a. ASHRAE Standard 90.1-2007 (with errata but without addenda).
   b. The Department of Energy Verification Protocol.
   c. A signed statement from the designer.
   d. International Performance Measurement and Verification Protocol.

4. Optimal IAQ performance can:
   a. Create savings in electrical bills.
   b. Improve the productivity and health of building occupants.
   c. Cause higher operation cost.
   d. Create higher rents.

5. What is the best way to control Environmental Tobacco Smoke (ETS)?
   a. Ban smoking inside the building.
   b. Place exterior smoking area at least 25 feet away from operable windows, entrances and air intakes.
   c. All interior spaces shall have negative pressure.
   d. Both a and b.

6. With regard to the Construction IAQ Management Plan, if your project uses permanently installed air handlers during construction, you should use
   a. Filtration media with a minimum Efficiency Reporting Value (MERV) of 6 at each return air grille.
   b. Filtration media with a minimum Efficiency Reporting Value (MERV) of 7 at each return air grille.
   c. Filtration media with a minimum Efficiency Reporting Value (MERV) of 8 at each return air grille.
   d. Filtration media with a minimum Efficiency Reporting Value (MERV) of 9 at each return air grille.

7. For Indoor Chemical and Pollutant Source Control, you need to
   a. Physically isolate activities related to chemical use.
   b. Submit drawings and cutsheets for the plumbing systems in chemical mixing area.
   c. Design the exterior sidewalk and pavement to drain away from the building at 2% minimum slope.
   d. None of the above.

8. Placing a lighting control in a hallway does not help you with Controllability of Systems because
   a. You need to make sure the hallway does not have a dead end corridor that is over 20 feet in length.
   b. You also need to add temperature control.
   c. You also need to make sure the hallway has a view to the outside.
   d. None of the above.

9. Which of the following standard(s) is mentioned with regard to Increased Ventilation?
   a. ASHRAE Standard 52.2-1999.
   b. ASHRAE Standard 62.1-2007.
   c. ASHRAE Standard 90.1-2007 (with errata but without addenda).
   d. SCAQMD Rule 1168.

10. Which of the following standard(s) is (are) mentioned in Light Pollution Reduction?
    a. ASHRAE/IESNA Standard 90.1-2007 (with errata but without addenda).
    b. IESNA RP-33.
    c. International Dark Sky Association Lighting Standard.
    d. Both a and b.

11. California Title 24-2005 is considered to be equal to ASHRAE/IESNA Standard 90.1-2007 (with errata but without addenda) for the following LEED-NC credit(s):
    a. Minimum Energy Performance – EAp2.
    b. Optimize Energy Performance – EAc1.
    c. Green Power-EAc6.
    d. All of the above.

12. Which of the following is not a rapidly renewable material?
    a. Bamboo flooring.
    b. Linoleum Flooring
    c. Glass Windows
    d. Wool Carpeting

13. What does "Xeriscape" mean?
    a. Drip Irrigation to save water.
    b. "Dry Landscape" design by using plants that use little or no water.
    c. Recycle existing plants on the project site.
    d. Reuse graywater for landscape irrigation.

14. Which of the following is a responsibility for the contractor to support the LEED documentation process?
    a. Document and provide calculations for waste diverted from landfill.
    b. Maintain a submittal log.
    c. Maintain a RFI log.
    d. Provide written documentation to justify a change order for rough grading.

15. Rapidly renewable material may also qualify for which credit(s)?
    a. MRc3: Material Reuse
    b. MRc5: Regional Materials
    c. MRc7: Certified Wood.
    d. All of the above.

16. Which of the following three statements are correct? (Choose 3)
    a. For non-residential projects, water closet uses per day per female FTE is 3.
    b. For residential projects, water closet uses per day per female is 5.
    c. The flow rate for a conventional water closet is 1.8 gpf.
    d. The flow rate for a low-flow water closet is 1.1 gpf.

17. Which of the following is the standard for Low-Emitting Sealants, Sealant Primers and Adhesives?
    a. SCAQMD (South Coast Air Quality Management District) Rule #1168, VOC limits effective 7/1/2005.
    b. Green Seal Standard for Commercial Adhesives GS-36 requirements with an effective date of 10/19/2000
    c. Green Label Plus
    d. Green Guard

18. Why should a developer locate a green building in a previously developed urban area?
    a. It is close to public transportation.
    b. It can use existing community services.
    c. It is close to existing utilities.
    d. a and b
    e. a, b, and c

19. Blackwater is water drained from a:
    a. kitchen sink.
    b. toilet.
    c. both a and b.
    d. None of the above

20. With regard to ozone depletion potential (ODP), the order from high ODP to low ODP is:
    a. CFC> HFC>HCFC
    b. CFC>HCFC>HFC
    c. HCFC>HFC>CFC
    d. HCFC> CFC>HFC

Exam Part Two sample questions:

21. A developer is building a new shopping center on a landfill site with a large amount of underground methane gas. Which three of the following statements are correct?
    a. The contractor can use perforated pipe under the building to collect the methane gas, and then vent it to a certain height above the roof.
    b. The contractor can use the "pump and treat" method to treat the methane gas.
    c. The contractor can use the "pump and treat" method to treat the groundwater.
    d. The contractor can use in-situ remediation.
    e. The contractor cannot vent the methane gas, and has to burn it at the site per EPA regulation.
    f. The contractor can use the solar detoxification strategy.

22. A store owner wants to rent a space in a base building that can satisfy SSc3.1: Alternative Transportation: Public Transportation Access. What requirements must the building meet to qualify for LEED point(s) under SSc3.1? (Choose 2)
    a. Locate within ¼-mile walking distance of an existing commuter rail.
    b. Locate within ½-mile walking distance of an existing commuter rail.
    c. Locate within ¼-mile walking distance of a bus stop.
    d. Locate within ¼-mile walking distance of one or more bus stops for 2 or more public bus.

23. A developer is building a commercial building. He is concerned about the cost related to IEQc6.1: Controllability of Systems: Lighting. How can the design team convince him about potential synergy and savings in first cost?
    a. The reduction in related heat load will reduce the size and amount of HVAC equipment.
    b. The lighting system will reduce tenants' electrical costs.
    c. IEQc6.1 will increase occupant comfort and in turn increase production.
    d. A lighting system qualifying for IEQc6.1 can be integrated with sensors and natural light to reduce energy costs.

24. A supermarket project has south-facing clear glazing. The design team is considering installing a tint film on the glazing to reduce HVAC heat load. Which of the following will achieve point(s) for IEQc8.1 and IEQc8.2: Daylight and Views?
    a. Using the tint film on the glazing as proposed, with 83% of the space having a view, and 73% of the space with access to daylight
    b. Using a shading device instead of the proposed film, with 88% of the space having a view, and 78% of the space with access to daylight
    c. Using a shading device instead of the film, with 95% of the space having a view, and 70% of the space with access to daylight
    d. Using a shading device instead of the film, with 95% of the space having a view, and 80% of the space with access to daylight

25. A contractor salvages doors and windows from an existing building. What LEED credits may he achieve? (Choose 2)
    a. Recycled Content
    b. Building Reuse
    c. Materials Reuse
    d. Construction and Demolition Waste
    e. Rapidly Renewable Materials

26. A new retail building has 200 occupants. What is the baseline potable water usage for sewage conveyance for WEc1: Water Use Reduction?
    a. 192,800 gallons (729,826 liters)
    b. 200,000 gallons (757,082 liters)
    c. 218,400 gallons (826,732 liters)
    d. 280,600 gallons (1,062,186 liters)

27. For the same retail building as described in Question 26 above, how much water can the design team save if they use waterless urinals and single flush, pressure assist HETs?
    a. 104,400 gallons (395,197 liters)
    b. 114,400 gallons (433,051 liters)
    c. 124,400 gallons (470,905 liters)
    d. 134,400 gallons (508,759 liters)

28. A developer is planning to spend 25 million dollars to build a new modern-style commercial building on a 12-acre landfill site. The buildings in the neighborhood are all Spanish style buildings. Which of the following statements is true regarding this building's LEED certification? (Choose 2)
    a. The base building design team needs to perform a Phase I Environmental Assessment (per ASTM E1527-05) to find out if the site is contaminated.
    b. The developer needs to perform a Phase II Environmental Assessment (per ASTM E1903-97, 2002) if the project team suspects the site is contaminated.
    c. This project is not eligible for LEED certification.
    d. The project can earn 1 point for SSc1, Ch2F1: Brownfield Redevelopment if the tenant's project team provides successful documentation for the remediation of the site.
    e. This project is not eligible for LEED certification because it is not compatible with the neighborhood building style.

29. The term "off-gassing" is related to:
    a. SSc3.1: Alternative Transportation: Public Transportation Access.
    b. SSc1, Ch2F5: Heat Island Effect: Roof.
    c. MRp1: Storage and Collection of Recyclables.
    d. IEQc4.5: Low-Emitting Materials: Systems Furniture and Seating.

30. CEE stands for:
    a. Consortium for Energy Efficiency.
    b. Consortium for Energy Evaluation.
    c. Committee for Energy Efficiency.
    d. Consumer Energy Efficiency.

31. NRC stands for:
    a. National Resource Center.
    b. Noise Reduction Coefficient.
    c. National Resource Consumption.
    d. Noise Reduction Coefficiency.

32. ETV stands for:
    a. Environmental Test Verification.
    b. Environmental Test Value.
    c. Environmental Technology Verification.
    d. Environmental Technology Value.

33. The typical background-sound level for classrooms is about 30 dBA, and a noisy cafeteria has an 80 dBA sound level. This means that:
    a. The sound level for the cafeteria is about twice the volume of the background noise level of the classroom.
    b. The sound level for the cafeteria is about thirty-two times the volume of the background noise level of the classroom.
    c. The sound level for the cafeteria cannot be compared with the volume of the background noise level of the classroom.
    d. None of the above

34. Select three of the following standards or codes that are related to the most LEED prerequisites and credits:
    a. ASHRAE
    b. ANSI
    c. DOE
    d. EPA
    e. ASTM
    f. DHS

35. Select three of the following project phases that are related to the most LEED prerequisites and credits:
    a. Pre-Design
    b. Schematic Design
    c. Design Development
    d. Construction Documents

e. Construction Administration
f. Occupation/Operation

36. When was LEED v3.0 launched?
    a. April 27, 2009
    b. May 27, 2009
    c. June 26, 2009
    d. July 1, 2009

37. With regard to synergies, SSc1, Ch2F2: Storm Water Design: Quantity Control, is related to which of the following credits? (Choose 2)
    a. SSc1, Ch2F3: Storm Water Design: Quality Control
    b. SSc1, Ch2F4: Heat Island Effect: Non-Roof
    c. SSc1, Ch2F5: Heat Island Effect: Roof
    d. SSc1, Ch2F9: Innovative Wastewater Technologies

38. With regard to synergies, SSc1, Ch2F4: Heat Island Effect: Non-Roof, could relate to which of the following credits? (Choose 3)
    a. SSc1, Ch2F2: Storm Water Design: Quantity Control
    b. SSc1, Ch2F3: Storm Water Design: Quality Control
    c. SSc1, Ch2F5: Heat Island Effect: Roof
    d. SSc1, Ch2F8: Water Efficient Landscaping: No Potable Water Use or No Irrigation
    e. IEQc7: Thermal Comfort

39. With regard to submittals, what does the project team need to submit for MRp1: Storage and Collection of Recyclables? (Choose 2)
    a. A tenant guideline
    b. Manufacturer's data sheets if available
    c. Floor plans, site plans, etc. if needed
    d. A recycling plan

40. With regard to project team coordination, who is responsible for submitting MRp1: Storage and Collection of Recyclables?
    a. The architect
    b. The developer
    c. The tenant
    d. a or c

II. Answers for the LEED AP ID+C Exam sample questions

Part One:

1. Answer: a
2. Answer: c
3. Answer: d
4. Answer: b
5. Answer: d
6. Answer: c
7. Answer: a
8. Answer: d
9. Answer: b
10. Answer: d
11. Answer: d
12. Answer: c
13. Answer: b
14. Answer: a
15. Answer: d
16. Answer: a, b and d
17. Answer: a
18. Answer: e
19. Answer: c
20. Answer: b

Part Two:

21. Answer: a, c and d
    Choice f is NOT correct and is dangerous for the project because of the large amount of underground methane gas at the site.
22. Answer: b and d
23. Answer: a. The key in this question is first cost. While other answers may be reasonable and can reduce long term cost, they cannot reduce first cost. The GBCI loves to use words like this to trick you. So, read the questions carefully, and do not get tricked by the GBCI.
24. Answer: d
25. Answer: c and d
26. Answer: c
    Assuming 50% male and 50% female
    5 work days per week x 52 weeks = 260 work days per year
    Refer to the following detailed calculation:

Fixture Type	**Daily Use**	**Flowrate (gpf)**	**Occupants**	**Sewage Generation (gal)**
WC (Male)	1	1.6	100	160
WC (Female)	3	1.6	100	480
Urinal (Male)	2	1.0	100	200
Urinal (Female)	0	1.0	100	0
Total Daily Volume (gal)				840
Annual Work Days				260
Total Annual Volume (gal)				218,400

27. Answer: a
Refer to the following detailed calculation:

Fixture Type	Daily Use	Flowrate (gpf)	Occupants	Sewage Generation (gal)
Single flush, pressure assist HET (Male)	1	1.0	100	100
Single flush, pressure assist HET (Female)	3	1.0	100	300
Waterless Urinal (Male)	2	0	100	0
Urinal (Female)	0	1.0	100	0
Total Daily Volume (gal)				400
Annual Work Days				260
Total Annual Volume (gal)				104,000
Total Annual Volume of Water Saved (gal)	218,400-104,000 = 114,400 Or 52.38%			

28. Answer: a and d. See SSc1, Ch2F1 for explanation.
29. Answer: d. See IEQc4.5 for explanation.
30. Answer: a. In the real LEED exam, you may run into some questions that you have not read about at all, this can be one of them. If you do not know the answer, you should simply use the guess strategy and use the same answer for guessing instead of spending a lot of time on the question.
31. Answer: b. In the real LEED exam, you may run into some questions that you have not read about at all, this can be one of them. If you do not know the answer, you should simply use the guess strategy and use the same answer for guessing instead of spending a lot of time on the question.
32. Answer: c. See IEQc4.3 for explanation.
33. Answer: b. One measurement used for acoustical performance is **A-weighted decibel (dBA) measurement** of noise level. It is a nonlinear scale: if you double the noise level, the measurement will increase 10 dBA instead of double. So, when the sound level increases from 30 dBA to 80 dBA, the volume of the sound increases 2 x 2 x 2 x 2 x 2 = 32 times. In the real LEED exam, you may run into some questions that you have not read about at all, this can be one of them. If you do not know the answer, you should simply use the guess strategy and use the same answer for guessing instead of spending a lot of time on the question.
34. Answer: a, b and d
35. Answer: a, b and c
36. Answer: a
37. Answer: a and b
38. Answer: a, b and d
39. Answer: c and d
40. Answer: d

III. LEED AP ID+C Mock Exam

Part One:

1. What is the maximum number of points that you can earn for LEED-CI?
   a. 50 points
   a. 60 points
   b. 80 points
   c. 110 points

2. What are the different levels of LEED-Schools Certification?
   a. LEED-SILVER, LEED-GOLD, and LEED-PLATINUM
   b. LEED-CERTIFIED, LEED-SILVER, LEED-GOLD, and LEED-PLATINUM
   c. LEED-BRONZE, LEED-SILVER, LEED-GOLD, and LEED-PLATINUM
   d.LEED-BRONZE, LEED-SILVER, and LEED-GOLD

3. How many categories are there for LEED-NC? What are they?
   a. There are 4. They are Sustainable Site Development (SS), Energy and Atmosphere (EA), Indoor Environmental Quality (IEQ), and Regional Priority (RP).
   b. There are 4. They are Sustainable Site Development (SS), Water Efficiency (WE), Energy and Atmosphere (EA), and Indoor Environmental Quality (IEQ).
   c. There are 6. They are Sustainable Site Development (SS), Water Efficiency (WE), Energy and Atmosphere (EA), Materials and Resource (MR), Indoor Environmental Quality (IEQ), and Regional Priority (RP).
   d. There are 7. They are Sustainable Site Development (SS), Water Efficiency (WE), Energy and Atmosphere (EA), Materials and Resource (MR), Indoor Environmental Quality (IEQ), Innovation and Design Process (ID), and Regional Priority (RP).

4. When was the LEED-NC 2.2 EA section revised?
   a. May 25, 2006
   b. July 26, 2006
   c. June 26, 2006
   d. March 23, 2006

5. What does LEED stand for?
   a. Leadership in Energy and Environmental Design.
   b. Leadership in Efficient Environmental Design.
   c. Leadership in Energy Efficiency Design.
   d. Leadership in Engineering and Environmental Design.

6. When was the USGBC founded?
   a. 1992
   b. 1993
   c. 1996
   d. 1998

7. Which of the following professionals is not involved in green building design?
   a. Architect
   b. Mechanical Engineer

c. Landscape Architect
d. Equipment Vendor

8. When was LEED-NC v2.2 published?
   a. 2003
   b. 2002
   c. 2001
   d. 2000

9. For LEED-NC, how many points are needed for Silver Certification?
   a. 30 points
   b. 40 points
   c. 50 points
   d. 60 points

10. For Construction Activity Pollution Prevention, ESC stands for:
    a. Environmental Safety Council
    b. Environment and Sedimentation Control
    c. Engineering Safety Council
    d. Erosion and Sedimentation Control

11. For Development Density and Community Connectivity, you need to construct a building:
    a. On a previously developed site AND within ¼ mile of 10 basic services AND within a neighborhood with 10 units per acre net on average or within a residential zone.
    b. On a previously developed site AND within one mile of 10 basic services AND within a neighborhood with 10 units per acre net on average or within a residential zone.
    c. On a previously developed site AND within ½ mile of 10 basic services AND within a neighborhood with 60 units per acre net on average or within a residential zone.
    d. On a previously developed site AND within ½ mile of 10 basic services AND within a neighborhood with 10 units per acre net on average or within a residential zone.

12. To get one point for Alternative Transportation, your building needs to be: (Choose 2)
    a. Located within ½ mile of a subway or rail station.
    b. Located within ¼ mile of two bus stops.
    c. Located within ¼ mile of one or more bus stops for two or more bus lines.
    d. Located within ½ mile of one or more bus stops for two or more bus lines.

13. You can get one point for Alternative Transportation by: (Choose 2)
    a. Providing secure bike storage and/or rack for 3% of the building users.
    b. Providing secure bike storage and/or rack for 5% of the building users.
    c. Providing secure bike storage and/or rack for 5% of the building users within 200 yards of the building entrance.
    d. Provide shower and changing facilities for 0.5% of FTE.

14. You can get three points for EAc1: Optimize Energy Performance: Lighting Power by:
    a. Improving a new building's energy performance by 10%.
    b. Improving an existing building's energy performance by 15%.
    c. Improving a new building's energy performance by 25%.
    d. Both a and b.

15. You can get eight points for Waste Water Reduction by:
    a. Reducing water use by 20%.
    b. Reducing water use by 25%.
    c. Reducing water use by 30%.
    d. Reducing water use by 35%.
    e. None of the above.

16. You can get at least one point for Regional Material if:
    a. At least 20% (based on cost) of the total materials have been manufactured within 500 miles of your project site.
    b. At least 10% (based on cost) of the total materials have been recovered, harvested, or extracted, and manufactured within 500 miles of your project site.
    c. At least 15% (based on cost) of the total materials have been manufactured within 500 miles of your project site.
    d. Both a and b

17. You can get one point for IEQ Category by:
    a. Providing daylight for 75% of the space.
    b. Providing daylight for 90% of the space.
    c. Providing a view for 75% of the space.
    d. Providing a view for 90% of the space.
    e. Either a or d.
    f. Either b or c.

18. What is the maximum number of points that you can get in the ID category?
    a. 6 points.
    b. 5 points.
    c. 4 points.
    d. 3 points.

19. If you have only one LEED AP on your team as a job captain, how many point(s) can you get?
    a. 0 points.
    b. 1 point.
    c. 2 points.
    d. 3 points.

20. ASHRAE means
    a. American Society of Heating, Refrigerating, Air Conditioning, and Environment.
    b. American Society of Heating, Refrigerating, Air Conditioning, and Engineering.
    c. American Society of Heating, Refrigerating, and Air Conditioning Engineers.
    d. None of above.

21. Which two of the following are not potable water per LEED?
    a. Reclaimed irrigation water.
    b. Water in detention pond.
    c. Water from municipal water system.
    d. Well water.

22. You are doing a tenant improvement project in a 50,000 s.f. building with five equal floors in an area with no zoning requirements. Your tenant space has 38,000 s.f. To get two points for SSc3.3: Alternative Transportation: Parking Availability, you need to provide: (Choose 2)
    a. Parking spaces above and beyond the minimum zoning requirements.
    b. Parking spaces meeting the minimum zoning requirements.
    c. Parking spaces do not exceed the minimum zoning requirements.
    d. 5% new parking spaces.

23. For Fundamental Commissioning of the Building Energy System, you are working on a project with 45,000 gross s.f. and you are looking for a CxA. Which of the following statements is correct?
    a. The CxA shall be independent of the project's design team.
    b. The CxA shall be independent of the project's construction management.
    c. Both a and b.
    d. The CxA can be a qualified person on the construction or design team with the necessary experience.

24. To get two points for On-Site Renewable Energy, you must:
    a. Provide 2% renewable energy (based on annual building energy cost).
    b. Provide 2.5% renewable energy (based on annual building energy cost).
    c. Provide 5% renewable energy (based on annual building energy cost).
    d. Provide 5.5% renewable energy (based on annual building energy cost).

25. For Low-Emitting Materials, VOC means:
    a. Volatile Organic Compounds.
    b. Volatile Organic Components.
    c. Volume of Circulation.
    d. Validity of Credit.

26. For Outdoor Air Delivery Monitoring, the monitoring locations of carbon dioxide concentrations in densely occupied space for mechanically ventilated spaces shall be:
    a. 2'-6" to 6'-6" above the floor.
    b. 3' to 6' above the floor.
    c. 3'-6" to 6'-6" above the floor.
    d. 4'-6" to 6'-6" above the floor.

27. To get one point for MRc1.2: Building Reuse: Maintain Interior Nonstructural Components, you should:
    a. Maintain 25% of existing floors, walls and ceiling systems.
    b. Maintain 30% of existing floors, walls and ceiling systems.
    c. Maintain 35% of existing floors, walls and ceiling systems.
    d. Maintain 40% of existing floors, walls and ceiling systems.

28. To get one point for MRc1.1: Tenant Space: Long Term Commitment, the tenant should:
    a. Promise to stay in the same location for 10 years.
    b. Promise to stay in the same location for 15 years.
    c. Sign a lease to stay in the same location for 10 years.
    d. Sign a lease to stay in the same location for 15 years.

29. To get two points for Construction Management, you should:
    a. Divert 50% of the project's total waste from the landfill waste stream in your construction management plan.
    b. Divert 75% of the project's total waste from the landfill waste stream in your construction management plan.
    c. Divert 80% of the project's total waste from the landfill waste stream in your construction management plan.
    d. Divert 90% of the project's total waste from the landfill waste stream in your construction management plan.

30. To get one point for Enhanced Refrigerant Management, you can:
    a. Choose not to use refrigerants.
    b. Choose HVAC&R and refrigerants that have no or reduced emission (below the maximum threshold) of compounds contributing to ozone depletion and global warming.
    c. Use small HVAC units and other cooling equipment containing less than 0.50 lbs. of refrigerant.
    d. All of above and you should not use ozone-depleting substances for your fire suppression systems.

31. What is the minimum renewable energy contract term for Green Power?
    a. One year.
    b. Two years.
    c. Three years.
    d. Four years.

32. If you reduce water use within your building by 20%, how many points can you get for Water Use Reduction?
    a. 0 points.
    b. 1 point.
    c. 2 points.
    d. None of above.

33. If you reduce your landscape irrigation water use by 20%, how many points can you get?
    a. 2 points.
    b. 1 point.
    c. 0 points.
    d. None of above.

34. If you reduce your potable water use for landscape irrigation by 75%, how many points can you get for Water Efficient Landscaping?
    a. 2 points.
    b. 1 point.
    c. 0 points.
    d. None of above.

35. Choose the official USGBC website(s): (Choose 2)
    a. www.USGBC.com
    b. www.LEED.org
    c. www.USGBC.org
    d. www.Greenbuild365.org

36. If you reuse, refurbish, or salvage 10% (based on value) of the building materials for a project, how many point(s) can you get for Material Reuse?
    a. 0 points.
    b. 1 point.
    c. 2 points.
    d. 3 points.

37. For MRc4, Recycled Content, recycled content is defined by the:
    a. USGBC
    b. ISO
    c. ICC
    d. ICBO

38. For MRc4, Recycled Content, recycled content is based on:
    a. Cost
    b. Volume
    c. Weight
    d. None of above

39. You can get two points for Regional Material for LEED CI by: (Choose 2)
    a. Using building materials manufactured (within 300 miles of your project site) for a minimum of 20% (based on cost) of the total amount of materials used on the project.
    b. Using building materials recovered, extracted, and manufactured regionally (within 500 miles of your project site) for a minimum of 10% (based on cost) of the total amount of materials used on the project.
    c. Using building materials manufactured regionally (within 500 miles of your project site) for a minimum of 20% (based on cost) of the total amount of materials used on the project.
    d. Using building materials manufactured regionally (within 500 miles of your project site) for a minimum of 30% (based on cost) of the total amount of materials used on the project.

40. For MRc7, Certified Wood is certified by the:
    a. USGBC
    b. FSC
    c. ICBO
    d. ISO

41. To get one point for Rapidly Renewable Materials for a CI project, you should:
    a. Use 2.5% of rapidly renewable materials based on cost.
    b. Use 5% of rapidly renewable materials based on cost.
    c. Use 7.5% of rapidly renewable materials based on cost.
    d. Use 10% of rapidly renewable materials based on cost.

42. SSc1: Site Selection requires you not to select a building located: (Choose 2)
    a. On prime farmland as defined by USDA.
    b. On virgin land with an elevation lower than five feet above the elevation of 100-year flood defined by FEMA.
    c. Within 200 feet of wetland defined by US Code of Federal Regulations 40 CFR.
    d. Either b or c

43. For SSc1, Ch2F3: Brownfield Redevelopment, a brownfield can be defined by: (Choose 2)
    a. ASTM
    b. Local, state or federal government agencies
    c. USGBC
    d. SCAQMD

44. With regard toSSc3.3 Alternative Transportation: Parking Availability, which of the followings are true: (Choose 2)
    a. Projects occupying less than 50% of the total building area have different requirements than larger projects.
    b. Projects occupying less than 75% of the total building area have different requirements than larger projects.
    c. Providing no new parking can gain two points for a project.
    d. You can gain a maximum of 3 points for this credit.

45. For Fundamental Refrigerant Management, you need to:
    a. Reduce the use of CFC-based refrigerants in new base building HVAC&R systems by 25%.
    b. Reduce the use of CFC-based refrigerants in new base building HVAC&R systems by 50%.
    c. Eliminate the use of CFC-based refrigerants in new base building HVAC&R systems.
    d. None of above.

46. For Enhanced Commissioning on a project over 100,000 s.f., the CxA shall:
    a. Not be an employee of the design firm.
    b. Not be an employee of the contractor holding construction contracts.
    c. Be a qualified employee of the owner.
    d. All of the above.

47. For Measurement and Verification, the M and V period shall cover:
    a. At least one year of post-construction occupancy.
    b. At least two years of post-construction occupancy.
    c. At least three years of post-construction occupancy.
    d. None of above.

48. Which of the following terms have been mentioned in EAc3: Measurement and Verification: (Choose 3)
    a. International Performance Measure and Verification Protocol (IPMVP).
    b. SCAQMD.
    c. ECMs.
    d. M&V.

49. If you increase the outdoor air ventilation rate by 20% above the rate required by ASHRAE, which of the following will improve:
    a. Heat island effect.
    b. Thermal comfort.
    c. Indoor air quality.
    d. Chance of satisfying EAp1: Fundamental Commissioning of the Building Systems.

50. You have to delete a horizontal window louver to save money for your project. Which of the following credits may be affected: (Choose 3)
    a. EAc1: Optimize Energy Performance.
    b. EAc4: Green Power.
    c. EAc2: Enhanced Commissioning.
    d. IEQc8.2: Daylight and Views.

51. Which of the following standards has a VOC limit for architectural paints, coatings, and primers applied for interior walls and ceilings?
    a. Bay Area Air Quality Management District Regulations 8.
    b. Green Seal Standard GS-11, Paints, First Edition.
    c. South Coast Air Quality Management District (SCAQMD) Rule 1113.
    d. International Performance Measure and Verification Protocol (IPMVP).

52. If your building meets the minimum requirements of Section 4 through 7 of ASHRAE 62.1-2007, Ventilation for Acceptable Indoor Air Quality (with errata but without addenda), you can get:
    a. 0 points for your project.
    b. 1 point for your project.
    c. 2 points for your project.
    d. 3 points for your project.

53. For mechanically ventilated space, to get one point for IEQc2, Increased Ventilation, you can:
    a. Increase the breathing zone outdoor air ventilation rates to all occupied spaces by at least 20% above the minimum rate required.
    b. Increase the breathing zone outdoor air ventilation rates to all occupied spaces by at least 30% above the minimum rate required.
    c. Increase the breathing zone outdoor air ventilation rates to all occupied spaces by at least 40% above the minimum rate required.
    d. Increase the breathing zone outdoor air ventilation rates to all occupied spaces by at least 50% above the minimum rate required.

54. Which of the following factors will affect thermal comfort?
    a. Air temperature, air speed, and relative humidity.
    b. Air temperature, air speed, floor to ceiling height, and relative humidity.
    c. Air temperature, radiant temperature, air speed, and relative humidity.
    d. None of above.

55. Which of the following will gain points for a project for both EAc1: Optimize Energy Performance and SSc1, Ch2F11: On-site Renewable Energy?
    a. Install horizontal window louvers.
    b. Use a reflective, light colored roof.
    c. Use better wall and roof insulation.
    d. Install active solar thermal systems using collection panels.

56. Your project involves renovation of a commercial building with seven floors each of equal area. You are doing renovation work on three floors and core and shell. Your scope of work includes mechanical and plumbing systems, tenant improvements, and exterior window replacement. Which of the following LEED rating systems shall you use?
    a. LEED-EB.
    b. LEED-EB and LEED-CS.
    c. LEED-CI and LEED-CS.
    d. LEED-EB and LEED-CS.

57. When should you start to prepare documentation for LEED credit submittals?
    a. In the pre-design stage.
    b. In the schematic design stage.
    c. At the beginning of the design development stage.
    d. Before construction.

58. What is the intent of IEQc6.1: Controllability of Systems: Lighting?
    a. To save electricity.
    b. To allow occupants the opportunity to customize their surroundings.
    c. To increase energy efficiency.
    d. To exceed ASHRAE requirements.

59. The intent of IEQc5: Indoor Chemical and Pollutant Source Control is:
    a. To avoid a chemical spill.
    b. To avoid building occupants' exposure to chemical pollutants and hazardous particles.
    c. Both a and b.
    d. None of above.

60. You are renovating a 20,000 s.f. building, and adding 60,000 s.f. of new steel-frame construction. You are using reclaimed wood from a local company and replacing single pane glazing with insulated glazing manufactured by a local firm. You may be able to get credit for the following categories: (Choose 3)
    a. MRc3.1: Materials Reuse.
    b. MRc5: Regional Materials.
    c. MRc2: Construction Waste Management.
    d. MRc7: Certified Wood.

61. Which two of the following statements are false?
    a. Your building cannot receive any credits at the design phase review.
    b. You can only file an appeal after the construction phase review.
    c. Your building can get LEED certification after the design phase review.
    d. If you do a design phase submittal and the project conditions change, you must submit additional information regarding the changes in the construction phase review.

62. Your LEED registered building has complicated issues and does not fall into any established credit categories or credit submittal templates. What should you do?
   a. Contact the USGBC team directly and try to resolve the issue.
   b. Pick a similar LEED credit category and submit documentation in this category.
   c. Provide a narrative explaining your complicated issues and describe how the credit intent is met.
   d. Follow the alternate compliance path option on the LEED submittal template and describe how your building complies.

63. Which of the following should not be in your specifications to the contractors for MRc2?
   a. The amount of waste leaving the site.
   b. A description of the waste materials.
   c. The requirements of a site logistic plan.
   d. The requirement of identifying the recyclers and haulers.

64. You design a sports club with 300 peak period occupants; you have included 10 bicycle racks inside the parking structure of the building. What must you do to get one point for Alternative Transportation?
   a. Increase the number of bicycle racks to 15.
   b. Place the bicycle racks inside a lockable, enclosed space.
   c. Make sure the bicycle racks are placed within 100 yards of the building.
   d. Add one shower and changing facility for each sex.

65. To get one point for MRc1.2: Building Reuse, what interior components can you reuse? (Choose 3)
   a. Columns
   b. Roof
   c. Window assemblies
   d. Building framing and skin
   e. Floor
   f. Ceiling

66. To get one point for SSc1, Ch2F4: Heat Island Effect: Non-Roof, you must select a base building that:
   a. Places 50% of the parking spaces under cover.
   b. Use paving materials with an SRI of 50 or higher.
   c. Provide an open grid pavement system for 50% of the parking surface.
   d. None of the above.

67. A base building has a roof with a slope of 1:12. To get one point for SSc1, Ch2F5: Heat Island Effect: Roof, it must:
   a. Use roofing material with a SRI of at least 29 for 75% of the roof surface.
   b. Use roofing material with a SRI of at least 78 for 75% of the roof surface.
   c. Provide a vegetated roof for at least 50% of the roof surface.
   d. Either b or c.
   e. Either a or c.

68. Select the incorrect statement(s) from the following options with regard to SSc1, Ch2F6: Light Pollution Reduction: (Choose 2)
    a. The zones of your projects are defined by IESNA RP-33.
    b. The zones of your projects are defined by ASHRAE.
    c. Both IESNA RP-33 and ASHRAE are related to this credit.
    d. LZ4 is the dark zone designation (Park and Rural Settings).

69. The imperviousness for a base building is 65%. To get one point for SSc1, Ch2F2: Storm Water Design: Quantity Control, it must:
    a. Apply a storm water management to reduce 25% of the volume of storm water runoff from the two-year 24-hour design storm.
    b. Apply a storm water management to reduce 50% of the volume of storm water runoff from the two-year 24-hour design storm.
    c. Apply a storm water management to reduce 30% of the volume of storm water runoff from the two-year, 24-hour design storm.
    d. None of the above.

70. The imperviousness for a base building is 55%. To get one point for SSc1, Ch2F3: Storm Water Design: Quality Control, it must:
    a. Have in place BMPs and site stormwater treatment systems designed to remove 60% of the average annual post-development total suspended solids (TSS) load per an existing monitoring report and 20% of the average annual total phosphorus (TP) of the site.
    b. Have in place BMPs and site stormwater treatment systems designed to remove 80% of the average annual post-development total suspended solids (TSS) load per an existing monitoring report and 40% of the average annual total phosphorus (TP) of the site.
    c. Have in place BMPs and site stormwater treatment systems designed to remove 90% of the average annual post-development total suspended solids (TSS) load per an existing monitoring report and 50% of the average annual total phosphorus (TP) of the site.
    d. None of the above.

71. Which of the following is not mentioned in IEQc3.1: Construction IAQ Management Plan?
    a. Sheet Metal and Air Conditioning Contractor Association (SMACNA)
    b. Minimum Efficiency Reporting Value (MERV)
    c. ASHRAE
    d. SCAQMD

72. You can satisfy IEQp2: Environmental Tobacco Smoke (ETS) Control by:
    a. Banning smoking in the building.
    b. Placing all outdoor designated smoking area a minimum of 25' from operable windows, outdoor air intakes and entries.
    c. Both a and b.
    d. None of the above.

73. To get one point for IEQc8.1: Daylight and Views, you can:
    a. Provide a daylight illumination level of at least 25 footcandles for at least 75% of the regularly occupied areas.
    b. Provide a daylight illumination level of at least 15 footcandles for at least 90% of the regularly occupied areas.
    c. Provide a daylight illumination level of at least 20 footcandles for at least 75% of the regularly occupied areas.
    d. None of the above.

74. To get one point for IEQc7.2: Thermal Comfort:
    a. Conduct a thermal comfort survey of building occupants within 6 months after occupants.
    b. Conduct a thermal comfort survey of building occupants within 12 months after occupants.
    c. Conduct a thermal comfort survey of building occupants within 18 months after occupants.
    d. Conduct a thermal comfort survey of building occupants within a period of 6 to 18 months after occupants.

75. The following terms are mentioned in IEQc4.4: Low-Emitting Materials: (Choose 3)
    a. Urea-formaldehyde resins.
    b. Medium Density Fiberboard (MDF).
    c. SCAQMD.
    d. Fit-out, Furniture, and Equipment (FF&E).
    e. ASHRAE

76. For IEQc3.2: Construction IAQ Management Plan, flush-out requires:
    a. Supplying 12,000 cu.ft. of outdoor air while maintaining humidity less than 60% and a minimum internal temperature of 60 degrees.
    b. Supplying 12,000 cu.ft. of outdoor air while maintaining humidity less than 80% and a minimum internal temperature of 60 degrees.
    c. Supplying 14,000 cu.ft. of outdoor air while maintaining humidity less than 60% and a minimum internal temperature of 60 degrees.
    d. Supplying 15,000 cu.ft. of outdoor air while maintaining humidity less than 60% and a minimum internal temperature of 60 degrees.

77. You are working on a new construction project and you provide an accessible area for storage and collection of recyclables. You will:
    a. Receive 0 points for the Materials and Resources LEED credit.
    b. Receive 1 point for the Materials and Resources LEED credit.
    c. Receive 2 points for the Materials and Resources LEED credit.
    d. None of the above.

78. Which of the following is false?
    a. Using a special kind of window to prevent birds from flying into the windows and getting killed may add one point to your project.
    b. You can get one extra ID point for WEc1: Water Use Reduction by saving 40% water.
    c. You need to have at least one LEED AP on your project team before project registration to get one point for IDc2: LEED Accredited Professional.
    d. No matter how many LEED APs your project team has, your project can get only one point for IDc2: LEED Accredited Professional.

79. Which of the following is not included in the commissioning report?
    a. Compliance with the design intent.
    b. Documentation for O&M.
    c. Specifications.
    d. Electrical riser diagram.

80. N.R.D.C. suggested that the US cease the production of CFCs in which year?
    a. 1985
    b. 1991
    c. 1993
    d. 1995

81. Which three of the following statements are correct? (Choose 3)
    a. A LEED project team leader is in charge of the LEED project check list.
    b. A LEED project team leader sets up LEED Online.
    c. A LEED project administrator has to be a LEED AP.
    d. A LEED project team assigns responsibilities to team members, and s/he is the only one who can input information to LEED Online
    e. A LEED project administrator is the point of contact between the LEED project team and the GBCI.

82. Which of the following statements are incorrect? (Choose 3)
    a. Only LEED project team members can search for existing CIRs in the GBCI/USGBC database.
    b. Only USGBC members can search for existing CIRs in the GBCI/USGBC database.
    c. The public can search for existing CIRs in the GBCI/USGBC database.
    d. The LEED project team members have to call the USGBC/GBCI for existing CIRs.

83. For LEED submittal phases: (Choose 3)
    a. You can only file appeals for a construction phase review.
    b. You can file appeals for design or construction phase reviews.
    c. The LEED project team decides which package to submit at the design phase.
    d. All WE credits can be submitted at the design phase.

84. Which one of the following is NOT addressed by SS of the LEED system?
    a. Noise pollution.
    b. Air pollution.
    c. Water pollution.
    d. CGP.

85. Stabilization techniques for a base building include: (Choose 3)
    a. Mulching.
    b. Silt fencing.
    c. Temporary seeding.
    d. Permanent seeding.

86. Which of the following are required submittals for SSc1, Ch2F2: Storm Water Design: Quantity Control?
    a. Storm quantities and rates for pre- and post-development situations.
    b. A narrative detailing the BMP, ESC, and responsible parties.
    c. A storm water plan assessment prepared by a civil engineer or other design professional.
    d. Reports or inspection logs, date-stamped photos, etc.
    e. All of the above.

87. For SS in the LEED system, wetlands are defined by the:
    a. FEMA
    b. USDA
    c. U.S. CFR
    d. Wetland Protection Act

88. For SS, basic services include, but are not limited to: (Choose 4)
    a. Banks
    b. Places of Worship
    c. Convenience Stores
    d. Bus stops
    e. Day Cares

89. Brownfields are defined by the:
    a. FEMA
    b. USDA
    c. U.S. CFR
    d. Wetland Protection Act
    e. EPA

90. Zero Emission Vehicles (ZEV) are defined by the:
    a. American Council for an Energy Efficient Economy (ACEEE)
    b. U.S. CFR
    c. EPA
    d. California Air Resources Board

91. Your project is at a previously developed site, and includes 55% landscaped area planted with adapted or native vegetation, a retention pond, a wetland, and a watercourse. The project's design strategies may contribute to: (Choose 3)
    a. SSc1, Ch2F7: Water Efficient Landscaping: Reduced by 50%
    b. WEc1: Water Use Reduction
    c. SSc1, Ch2F2: Storm Water Design: Quantity Control
    d. SSc1, Ch2F4: Heat Island Effect: Non-Roof
    e. IEQc2: Increased Ventilation

92. Natural and mechanical ventilation in a building may contribute to: (Choose 3)
    a. IEQc2: Increased Ventilation
    b. IEQp1: Minimum IAQ Performance
    c. EAp1: Fundamental Commissioning of the Building Energy Systems
    d. SSc1: Site Selection
    e. All of the above

93. Increasing graywater use, increasing rainwater harvesting and decreasing the demand on local water aquifers may contribute to:
    a. SSc1, Ch2F7: Water Efficient Landscaping: Reduced by 50%
    b. SSc1, Ch2F8: Water Efficient Landscaping: No Potable Water Use or No Irrigation
    c. WEc1, Water Use Reduction
    d. All of the above

94. Which of the following statements are correct? (Choose 2)
    a. Water use for a conventional water closet is 1.2 gpf.
    b. Water use for a conventional water closet is 1.6 gpf.
    c. The water use baseline for a residential lavatory is 2.5 gpm at 60 psi.
    d. The water use baseline for a public lavatory is 0.5 gpm at 60 psi.

95. Which of the following statements regarding water use for a water closet are correct? (Choose 2)
    a. 1.6 gpf for baseline
    b. 1.0 gpf for low-flow
    c. 1.1 gpf for HET dual flush (low-flush)
    d. 0.5 gpf for HET, foam flush

96. Which of the following statements are correct? (Choose 3)
    a. WaterSense is developed per EPA and ASHRAE standards.
    b. WaterSense is a partnership program sponsored by the EPA.
    c. WaterSense Standards exceed IPC and UPC requirements in some cases.
    d. Use of WaterSense fixtures may qualify your project for LEED credits.

97. The Energy Policy Act of 1992 is related to:
    a. WE
    b. SS
    c. EA
    d. IEQ
    e. ID

98. Which three of the following statements are correct?
    a. Showers are considered to be a 480 second duration for residential projects, and a 280 second duration for non-residential projects.
    b. High-efficiency (HE) fixtures include: single gravity, high-efficiency toilets (HET) and high-efficiency urinals (HEU).
    c. The flow rate for ultra low-flow lavatory use is 0.5 gpm.
    d. The flow rate for showers is 2.5 gpm at 60 psi per shower stall.
    e. The flow rate for a kitchen sink is 2.2 gpm at 60 psi.

99. A project has 100 tons of demolition waste. 30 tons will be donated, 20 tons will be re-used on-site, and 25 tons will be recycled. Which two of the following statements are correct?
    a. The project can get at least one point for MRc2: Construction Waste Management.
    b. The 30 tons of recycled material can be included in MRc4: Recycled Content.
    c. Since the donated materials cannot be included in the diverted waste, this project cannot gain any point for MRc2: Construction Waste Management.
    d. You cannot count the 20 tons of re-used materials toward MRc3: Materials Reuse if you already included the material in MR Credit 2, Construction Waste Management.

100. MEP equipment can be included in:
    a. MRc1.2: Building Reuse
    b. MRc2: Construction Waste Management
    c. MRc3: Materials Reuse
    d. MRc4: Recycled Content

Part Two:

101. With regard to submittals, what does a project team need to submit for MRc2: Construction Waste Management? (Choose 3)
   a. A list of actual materials costs, not including equipment and labor, for materials per CSI MasterFormat™ 2004 Edition Divisions 03-10, 31 (section 31.60.00 Foundations) and 32 (section 32.10.00 Paving, 32.30.00 Site Improvements, and 32.90.00 Planting), including Division 12 is optional
   b. A list of actual materials costs, including equipment and labor, for materials per CSI MasterFormat™ 2004 Edition Divisions 03-10, 31 (section 31.60.00 Foundations) and 32 (section 32.10.00 Paving, 32.30.00 Site Improvements, and 32.90.00 Planting), including Division 12 is optional
   c. A list of manufacturers' names, product names, costs, and percentage of pre- and post-consumer content
   d. Manufacturers' letters or cutsheets indicating the listed products' recycled content
   e. A plan showing the location of the recycle area

102. With regard to submittals, what does the project team need to submit for MRc5: Regional Materials? (Choose 3)
   a. A list of actual materials costs, not including equipment and labor, for materials per CSI MasterFormat™ 2004 Edition Divisions 03-10, 31 (section 31.60.00 Foundations) and 32 (section 32.10.00 Paving, 32.30.00 Site Improvements, and 32.90.00 Planting), including Division 12 is optional
   b. A list of actual materials costs, including equipment and labor, for materials per CSI MasterFormat™ 2004 Edition Divisions 03-10, 31 (section 31.60.00 Foundations) and 32 (section 32.10.00 Paving, 32.30.00 Site Improvements, and 32.90.00 Planting), including Division 12 is optional
   c. A list of product purchases extracted, harvested, or manufactured regionally.
   d. Paperwork showing manufacturers' names, product names, costs, the distance between the extraction site and the project, and the distance between the manufacturer and the project
   e. Manufacturers' letters or cutsheets indicating the materials' origin and manufacture within a 300 miles radius of the site

103. With regard to IEQc5: Indoor Chemical and Pollutant Source Control, which of the following rooms shall be isolated, i.e., the areas are exhausted sufficiently to create a negative pressure when the doors to these areas are closed, provide hard lid ceilings, deck-to-deck partitions, and self-closing doors, etc? (Choose 3)
   a. Printing rooms
   b. Laundry rooms
   c. Gift shops
   d. Public restrooms
   e. Science laboratories

104. You are working on a LEED CI project with 20,000 s.f. of tenant space in a 50,000 s.f. base building. The local codes require 5 parking spaces per 1,000 s.f. Which of the following statements are true with regard to SSc3.3: Alternative Transportation: Parking Availability? (Choose 2)
   a. The number of parking required for the tenant space is 100.
   b. The number of parking required for the tenant space is 250.
   c. You need to provide 5 preferred parking spaces for carpools or vanpools.
   d. You need to provide 15 preferred parking spaces for carpools or vanpools.
   e. The number of preferred parking spaces for carpools or vanpools cannot be determined.

105. A developer is building a new retail building with 12,000 s.f. of gross leasable area. He intends to use 2000 s.f. as a leasing office and lease the remaining space to various shop tenants. Which LEED building rating system should he choose?
   a. LEED NC
   b. LEED CS
   c. LEED ND
   d. Either a or b
   e. None of the above

106. The US Army is building a 120,000 s.f. barrack, the building to site ratio is 1.5%. Which LEED building rating system should the project team choose?
   a. LEED NC
   b. LEED CS
   c. LEED ND
   d. This project cannot seek LEED certification.
   e. None of the above

107. For a LEED CS project, which of the following statements is true?
   a. All LEED project team members have access to the CIR database and LEED Online after registration.
   b. The LEED project team may need to report post-occupancy water and energy use.
   c. The GBCI will inform the project team credit is awarded or denied after design submittal.
   d. None of the above

108. For a LEED for Homes project, which of the following statements is true?
   a. Homebuilders who intend to achieve LEED for Homes certification must contact a LEED for Homes provider organization.
   b. Homebuilders who intend to achieve LEED for Homes certification must contact a LEED Green Rater.
   c. A LEED AP carries out the thermal bypass inspection.
   d. None of the above

109. Which of the following statements regarding LEED are true? (Choose 3)
    a. CIR stands for Credit Interpretation Rulings.
    b. Precertification is for LEED CS or LEED NC.
    c. The USGBC will start to review your project 10 working days after receiving your submittal.
    d. A Green Rater is a professional who has shown qualifications for an energy rating certification, and demonstrates abilities to deal with a home's energy systems and performance.
    e. With regard to LEED Certification, you have 20 days to appeal any comments from the USGBC. You need to pay $500 for each appeal.

110. Which of the following statements regarding CIRs are true? (Choose 2)
    a. If a LEED AP is not sure about where to place the handicap parking stalls, he can seek advice from the GBCI.
    b. One advantage of using a CIR is that it will let you know for sure if you can achieve a credit.
    c. Always review the GBCI and USGBC websites for previous CIRs before you submit one.
    d. If you cannot find a similar CIR, then submit a new CIR to GBCI online.

111. Which of the following statements regarding CIRs are true? (Choose 2)
    a. Be sure to include your contact information and the name of the credit so that the GBCI can respond to you in a prompt manner.
    b. Check your projects against each LEED credit or prerequisite before submitting a CIR.
    c. Check the USGBC reference guide before submitting a CIR.
    d. Attach site plans, floor plans, and any other related drawings in your CIR to help clarify the issues if these plans are available.

112. Which of the following statements are not true? (Choose 2)
    a. The separation of the tasks for the USGBC and the GBCI is to meet the protocols of ASTM and ISO.
    b. To qualify for LEED NC, a building must have a minimum area of 1,000 s.f.
    c. A portable classroom building may qualify for LEED NC or LEED Schools.
    d. A LEED project has to comply with environmental laws.

113. A fast food restaurant has leased a 950 s.f. space and is doing tenant improvement work for the entire space. Which of the following statements are true? (Choose 2)
    a. This project must meet ADA guidelines.
    b. This project can seek LEED NC certification.
    c. This project can seek LEED CI certification.
    d. This project cannot seek LEED certification because it is too small.

114. Which of the following statements are not true? (Choose 2)
    a. Green buildings employ an integrative approach and encourage the participation of key stake holders.
    b. A LEED project team may include facilities management staff, facilities O&M staff, community members, the owner, and planning staff.
    c. A LEED project team does not include product manufacturers.
    d. A LEED project team does not include structural engineers.

115. What do you need to submit to achieve SSc1, Ch2F9: Innovative Wastewater Technologies? (Choose 3)
   a. Calculation of the number and type of occupants.
   b. A narrative detailing the BMP, ESC, and responsible parties.
   c. Information on system schematics as well as the capacity of graywater and rainwater systems
   d. Field reports or shop drawing logs, product cut sheets, etc.
   e. Manufacturer's data showing manufacturer, model, and water consumption rate of each fitting and fixture

116. SSc1, Ch2F9: Innovative Wastewater Technologies may affect the following prerequisite or credits: (Choose 3)
   a. SSc1: Site Selection
   b. SSc1, Ch2F4: Heat Island Effect: Non-Roof
   c. SSc1, Ch2F7: Water Efficient Landscaping: Reduced by 50%
   d. EAc2: Enhanced Commissioning
   e. EAc3: Measurement and Verification

117. SSc2: Development Density and Community Connectivity may contribute to which of the following prerequisites or credits: (Choose 2)
   a. WEp1: Water Use Reduction
   b. SSc1: Site Selection
   c. SSc3.1: Alternative Transportation: Public Transportation Access
   d. EAc1: Outdoor Air Delivery Monitoring

118. Which of the following can be included as basic service for SSc2: Development Density and Community Connectivity??
   a. Wachovia Bank
   b. Pet Hospital
   c. Sports Club
   d. SAT Prep Academy
   e. All of the above

119. A developer is doing a major remodel project. The LEED project team can include the existing MEP equipment in:
   a. MRp1: Storage and Collection of Recyclables
   b. MRc1 for CS: Building Reuse: Maintain Existing Walls, Floors, and Roof
   c. MRc2: Construction Waste Management
   d. MRc3: Materials Reuse

120. To achieve 2 points for MRc3: Materials Reuse, a project needs to have:
    a. A total of 10% or more (based on cost) of the materials used be reused, refurbished, or salvaged materials. The approved materials can be from off-site sources.
    b. A total of 10% or more (based on weight) of the materials used be reused, refurbished, or salvaged materials. The approved materials cannot be from off-site sources.
    c. A total of 20% or more (based on cost) of the materials used be reused, refurbished, or salvaged materials. The approved materials can be from off-site sources.
    d. A total of 20% or more (based on weight) of the materials used be reused, refurbished, or salvaged materials. The approved materials cannot be from off-site sources.

121. Which of the following statements is not true?
    a. MRc3: Materials Reuse is related to most of the credits in the MR category.
    b. MRc3: Materials Reuse is based on weight.
    c. MRc3: Materials Reuse is based on cost.
    d. The submittals for MRc3: Materials Reuse can be handled by the architect.

122. Which of the following statements are not true with regard to MRc6: Rapidly Renewable Materials? (Choose 2)
    a. Rapidly renewable materials are those produced from plants that can be generated in less than a 10-year lifecycle.
    b. Rapidly renewable materials are those produced from plants that can be generated in less than a 15-year lifecycle.
    c. If at least 5% (based on cost) of the building products' and materials' total value for your job are rapidly renewable materials, your team can achieve 1 point for LEED NC.
    d. If at least 5% (based on weight) of the building products' and materials' total value for your job are rapidly renewable materials, your team can achieve 1 point for LEED CS.

123. To achieve IEQp2: Environmental Tobacco Smoke (ETS) Control, a LEED project team needs to submit the following: (Choose 2)
    a. Manufacturer's data sheets showing the air-tight properties of the partition.
    b. An ETS policy describing non-smoking areas.
    c. Test results for air quality in non smoking areas.
    d. A record of testing data for interior smoking rooms to show that there is no cross contamination.

124. Who is the responsible party for SSc1, Ch2 F1: Brownfield Redevelopment?
    a. The Contractor
    b. The Civil Engineer, Environmental Engineer, or Geotechnical Engineer
    c. The Plumbing Engineer
    d. The tenant
    e. Either b or d

125. SSc1, Ch2 F1: Brownfield Redevelopment is related to the following? (Choose 3)
    a. ASTM
    b. ANSI
    c. ASHRAE
    d. CERLA
    e. U.S. EPA

126. To achieve SSc3.1: Alternative Transportation: Public Transportation Access, the project may need to meet the following requirements? (Choose 2)
   a. The project shall be located within ¼ mile of one planned and funded, or one existing, commuter rail or subway station.
   b. The project shall be located within ½ mile of one planned and funded, or one existing, commuter rail or subway station.
   c. The project shall be located within ¼ mile of at least one stop for two or more campus bus lines usable by the occupants of the building.
   d. All distances above shall be measured from a main building entrance.

127. Which of the following statements is not true with regard to SSc3.1: Alternative Transportation: Public Transportation Access? (Choose 3)
   a. A project team can achieve a maximum of 3 points for this credit.
   b. A project team can achieve a maximum of 4 points for this credit.
   c. A project team can achieve a maximum of 6 points for this credit.
   d. None of the above

128. Which of the following statements are not true with regard to SSc1, Ch2F8: Water Efficient Landscaping: No Potable Water Use or No Irrigation? (Choose 3)
   a. A project team can achieve 2 points for this credit.
   b. A project team can achieve 4 points for this credit.
   c. A project team cannot recycle wastewater because this credit is related to outdoor water.
   d. None of the above

129. Which of the following statements are true with regard to SSc1, Ch2F9: Innovative Wastewater Technologies? (Choose 2)
   a. WEc2 is applicable to outdoor water.
   b. WEc2 is applicable to indoor water.
   c. A project team can achieve 2 points for SSc1, Ch2F9
   d. A project team can achieve 4 points for SSc1, Ch2F9

130. Which of the following statements are not true with regard to SSc1, Ch2F9: Innovative Wastewater Technologies? (Choose 2)
   a. SSc1, Ch2F9 is related to all other prerequisites or credits concerning water use.
   b. SSc1, Ch2F9 is related to most of other prerequisites or credits concerning water use.
   c. A project team can use a living machine/natural system.
   d. A project team cannot use a packaged mechanical wastewater treatment system.

131. Which of the following statements are true with regard to EAp1: Fundamental Commissioning of the Building Energy Systems? (Choose 2)
   a. The project team may need to finish the commissioning process activities for daylighting and lighting controls.
   b. The project team may need to finish the commissioning process activities for mechanical and passive HVAC and refrigeration systems and related controls
   c. The project team shall finish the commissioning process activities for wind, solar, or other renewable energy systems.
   d. The project team shall finish the commissioning process activities for domestic hot water systems

132. Which of the following are examples of process energy uses? (Choose 2)
    a. HVAC
    b. Refrigeration and kitchen cooking
    c. Space heating
    d. Elevators and escalators

133. Which of the following are examples of regulated (non-process) energy uses? (Choose 2)
    a. Exhaust fans and hoods
    b. Office and general miscellaneous equipment
    c. Lighting as part of the medical equipment
    d. Service water heating

134. Which of the following statements are not true with regard to EAc1.3: Optimize Energy Performance: HVAC? (Choose 2)
    a. A project team can only use the Prescriptive Compliance Method to achieve points.
    b. A project team can only use the Whole Building Energy Simulation Method to achieve points.
    c. A project team can achieve 8 points for EAc1.3.
    d. A project team can achieve 10 points for EAc1.3.

135. A project team is working on an office building, and the team wants to pursue LEED NC certification under LEED v3.0. The team has reviewed the LEED NC scorecard and determined the maximum points the project can get are: 9 points for SS, 11 points for WE, 10 points for EA, 8 points for MR, 10 points for IEQ. What is the highest level of building LEED certification they can achieve for this project?
    a. Certified
    b. Silver
    c. Gold
    d. Platinum

136. SSc1, Ch2F11: On-Site Renewable Energy is typically handled at which of the following project phases?
    a. Pre-Design
    b. Schematic Design
    c. Design Development
    d. Construction Documents
    e. Construction Administration
    f. Occupation/Operation

137. EAc1.1: Optimize Energy Performance: Lighting Power is typically not handled at which of the following LEED submittal phases? (Choose 3)
    a. Design
    b. Design Development
    c. Construction
    d. Construction Administration

138. A project team is working on a new building, and the team wants to pursue LEED CS certification under LEED v3.0. The team has reviewed the LEED NC scorecard and determined the maximum points the project can get are: 7 points for SS, 10 points for WE, 8 points for EA,

6 points for MR, 8 points for IEQ. What is the highest level of building LEED certification they can achieve for this project?
a. Certified
b. Silver
c. Gold
d. Platinum

139. With regard to SSc1, Ch2F11: On-Site Renewable Energy, which of the following are eligible on-site systems? (Choose 2)
a. Wind energy systems
b. Geo-exchange systems
c. Architectural features
d. Geothermal heating systems

140. With regard to SSc1, Ch2F11: On-Site Renewable Energy, which of the following are ineligible on-site systems? (Choose 2)
a. Wave and tidal systems
b. Biofuel-base electrical systems
c. Passive solar
d. Daylighting

141. With regard to EAp3: Fundamental Refrigerant Management, which of the following are incorrect? (Choose 2)
a. A project may achieve 0 points for EAp3.
b. A project may achieve 1 point for EAp3.
c. A project may use 5% or less CFC content in its HVAC system.
d. A project cannot use any CFCs in its HVAC system.

142. With regard to EAp3: Fundamental Refrigerant Management, which of the following are correct? (Choose 2)
a. Small water coolers, small HVAC units, standard refrigerators, and other cooling equipment containing less than 1.5 lbs. of refrigerant are not included as part of the base building system and are exempt from its requirements.
b. EAp3 is related to IEQ credits.
c. You should use no CFC-based refrigerants for new buildings.
d. HVAC units with payback times longer than 10-years are feasible units.
e. HVAC units with payback times longer than 10-years are non-feasible units.

143. With regard to EAc3: Measurement and Verification, which of the following are correct? (Choose 3)
a. The project team may need to submit a guideline on tenant's energy use, including information on how to determine energy use and the related costs.
b. The project team may not need to submit a guideline on tenant's energy use, including information on how to determine energy use and the related cost.
c. The project team may not need an IPMVP-compliant M&V plan.
d. The project team needs to show and update the locations of any meters for measurement on a diagram.
e. The project team may not need to show and update the locations of any meters for measurement on a diagram.

144. Calibrated Simulation is based on:
    a. ASTM
    b. ANSI
    c. International Performance Measure and Verification Protocol (IPMVP)
    d. ASHRAE

145. EAc4: Green Power is related to: (Choose 2)
    a. Green-e Product Certification
    b. Green Label Plus
    c. Center for Resource Solution (CRS)
    d. ASHRAE

146. Which of the following are incorrect? (Choose 2)
    a. Paper, corrugated cardboard, metals, glass, and plastics make up about 60% of the total municipal solid waste stream.
    b. Based on percentage of weight, the remaining solid waste's ranking from highest to lowest is wood, food scraps, yard trimmings, textiles, leather, and rubber and other.
    c. The GBCI encourages you to compost food scraps and yard trimmings of waste on-site if possible.
    d. Yard trimmings cannot be composted on-site

147. Which of the following are true with regard to MRc3.1: Materials Reuse? (Choose 2)
    a. The salvaged and reused materials are calculated per their weight.
    b. The salvaged and reused materials are calculated per their cost.
    c. MRc3.1 makes reference to ASTM.
    d. MRc3.1 makes reference to the CSI MasterFormat™.

148. MRc4: Recycled Content is related to the following credits: (Choose 2)
    a. MRc1.2: Building Reuse
    b. MRc2: Construction Waste Management
    c. MRc3: Materials Reuse
    d. MRc5: Regional Materials

149. You can get one innovation point for MRc4: Recycled Content if:
    a. 10% of the materials are recycled.
    b. 20% of the materials are recycled.
    c. 30% of the materials are recycled.
    d. None of the above

150. MRc4: Recycled Content is related to the following:
    a. ISO
    b. ASTM
    c. ASHRAE
    d. EPA

151. Which of the following are incorrect? (Choose 2)
   a. A maximum of 2 points can be awarded for MRc5: Regional Materials.
   b. A maximum of 3 points can be awarded for MRc5: Regional Materials.
   c. A maximum of 1 point can be awarded for MRc6: Rapidly Renewable Materials.
   d. A maximum of 2 points can be awarded for MRc6: Rapidly Renewable Materials.

152. MRc3.1: Materials Reuse is typically handled at which of the following project phases?
   a. Pre-Design
   b. Schematic Design
   c. Design Development
   d. Construction Documents
   e. Construction Administration
   f. Occupation/Operation

153. Which of the followings are pre-consumer materials? (Choose 2)
   a. Paper
   b. Glass
   c. Materials from regrind
   d. Materials from rework

154. Which of the following are included in the LEED submittal package for MRc4: Recycled Content? (Choose 2)
   a. A list of actual materials costs
   b. A tenant guideline
   c. A list of manufacturers' names and manufacturers' letters or cutsheets
   d. The percentage of recycled materials by volume

155. A contractor is working on a LEED NC project. He needs to use 30 tons of material for the project, including 2 tons of various post-consumer recycled content and 2 tons of various pre-consumer recycled content. How many points can he achieve for MRc4: Recycled Content?
   a. 1 point
   b. 2 points
   c. There is not enough information to determine how many points he can achieve.
   d. None of the above

156. A contractor is working on a LEED NC project. He spent 30 million dollars on materials for the project, including 3 million dollars on various post-consumer recycled content and 5 million dollars on various pre-consumer recycled content. How many points can he achieve for MRc4: Recycled Content?
   a. 1 point
   b. 2 points
   c. There is not enough information to determine how many points he can achieve.
   d. None of the above

157. A contractor is working on a LEED NC project. He needs to use 30 tons of materials for the project, including 6 tons of materials manufactured regionally. How many points can he achieve for MRc5: Regional Materials?
    a. 1 point
    b. 2 points
    c. There is not enough information to determine how many points he can achieve.
    d. None of the above

158. A contractor is working on a LEED NC project. He spent 30 million dollars on materials for the project, including 6 million dollars on materials manufactured regionally. How many points can he achieve for MRc5: Regional Materials?
    a. 1 point
    b. 2 points
    c. There is not enough information to determine how many points he can achieve.
    d. None of the above

159. A contractor is working on a LEED NC project. He needs to use 30 tons of materials for the project, including 1 ton of rapidly renewable materials. How many points can he achieve for MRc6: Rapidly Renewable Materials?
    a. 0 points
    b. 1 point
    c. 2 points
    d. There is not enough information to determine how many points he can achieve.

160. A contractor is working on a LEED NC project. He spent 30 million dollars on materials for the project, including 5 million dollars on rapidly renewable materials. How many points can he achieve for MRc6: Rapidly Renewable Materials ?
    a. 0 points
    b. 1 point
    c. 2 points
    d. There is not enough information to determine how many points he can achieve.

161. With regard to synergies, IEQp1: Minimum IAQ Performance is related to which of the following? (Choose 2)
    a. WEp1: Water Use Reduction
    b. SSc1, Ch2F1: Brownfield Redevelopment
    c. SSc3: Alternative Transportation
    d. MRc2: Construction Waste Management

162. With regard to IEQp1: Minimum IAQ Performance, a LEED project team needs to submit which of the following?
    a. Calculations to confirm compliance with ASHRAE 62.1-2007
    b. A description of future tenants, expected uses and spaces types
    c. A tenant guideline
    d. A report showing the result of measurement and verification

163. What is the LEED Submittal Phase for IEQp1: Minimum IAQ Performance?
   a. Schematic Design
   b. Design
   c. Construction
   d. None of the above

164. IEQp2: Environmental Tobacco Smoke (ETS) Control is typically handled at which of the following project phases?
   a. Pre-Design
   b. Schematic Design
   c. Design Development
   d. Construction Documents
   e. Construction Administration
   f. Occupation/Operation

165. With regard to synergies, IEQp2: Environmental Tobacco Smoke (ETS) Control is related to which of the following? (Choose 2)
   a. SSc1, Ch2F1: Brownfield Redevelopment
   b. SSc1, Ch2F4: Heat Island Effect: Nonroof
   c. EAc1: Optimize Energy Performance
   d. EAc3: Enhanced Commissioning

166. Which of the following are not the responsible parties for IEQc2: Increased Ventilation? (Choose 2)
   a. The Contractor
   b. The Civil Engineer, Environmental Engineer, or Geotechnical Engineer
   c. Mechanical Engineer
   d. Architect

167. A LEED project team uses fly ash in concrete mixes. Which of the following statements are true? (Choose 2)
   a. The LEED project team may seek an ID point because this is a reuse of pre-consumer content.
   b. The LEED project team cannot seek an ID point because this is already covered under an existing LEED credit.
   c. The LEED project team cannot seek an ID point because fly ash presents environmental challenges in reuse.
   d. The design team needs to get the owner's consent for using the fly ash.

168. IEQc2: Increased Ventilation is related to which of the following? (Choose 3)
   a. ANSI
   b. ASHRAE
   c. ASTM Chartered Institution of Building Service Engineers (CIBSE)
   d. Carbon Trust "Good Practice Guide 237"
   e. International Performance Measure and Verification Protocol (IPMVP)

169. With regard to synergies, IEQc2: Increased Ventilation is related to which of the following? (Choose 3)
   a. EAp1: Fundamental Commissioning of the Building Energy Systems
   b. EAc2: Enhanced Commission
   c. EAc3: Measurement and Verification
   d. EAc4: Green Power
   e. IEQc5: Indoor Chemical and Pollutant Source Control

170. With regard to IEQc3.2: Construction IAQ Management Plan: Before Occupancy, which of the following are not true?
   a. A project team can seek 1 point for IEQc3.2.
   b. A project team can seek 2 points for IEQc3.2.
   c. IEQc3.2 can award one ID bonus point.
   d. IEQc3.2 does not apply to CI.

171. A project team is working on a LEED CI project. Which of the following can qualify as a basic service for SSc2: Development Density and Community Connectivity? (Choose 3)
   a. A Chinese restaurant
   b. A law office
   c. An Olive Garden restaurant
   d. A public restroom
   e. A Ralph's Supermarket

172. Which of the following is true with regard to SSc2: Development Density and Community Connectivity?
   a. A project team can achieve a maximum of 3 points.
   b. A project team can achieve a maximum of 4 points.
   c. A project team can achieve a maximum of 5 points.
   d. A project team can achieve a maximum of 6 points.

173. Which of the following are true with regard to IEQc3.2: Construction IAQ Management Plan: Before Occupancy? (Choose 2)
   a. After construction and all interior finishes are completed but before occupancy, the project team may need to do a flush-out by keeping the relative humidity less than or equal to 60%, and a minimum of 68 degrees Fahrenheit internal temperature, and provide a total air volume of 14,000 cubic feet (c.f.) of outdoor air per s.f. of floor area.
   b. A project team may need to do a baseline IAQ test after construction
   c. The project team can start the flush-out before completion of cleaning, the punch list, and final test of the control systems.
   d. The number of ventilation systems and size of the building dictate the number of sampling locations.

174. With regard to synergies, IEQc3.2: Construction IAQ Management Plan: Before Occupancy, is related to which of the following? (Choose 2)
   a. EAp1: Fundamental Commissioning of the Building Energy Systems
   b. EAc4: Enhanced Refrigerant Management
   c. IEQc4: Low-Emitting Materials
   d. IEQc5: Indoor Chemical and Pollutant Source Control

175. IEQc3.2: Construction IAQ Management Plan: Before Occupancy, is related to which of the following?
   a. ANSI
   b. ASHRAE
   c. ASTM Chartered Institution of Building Service Engineers (CIBSE)
   d. EPA

176. Which of the following is true with regard to IEQc4.1: Low-Emitting Materials: Adhesives and Sealants? (Choose 2)
   a. IEQc4.1 is related to SCAQMD (South Coast Air Quality Management District) Rule #1168.
   b. A project team can achieve a maximum of 2 points
   c. IEQc4.1 is related to Green Seal Standard for Commercial Adhesives GS-33 requirements with an effective date of 10/19/2000.
   d. IEQc4.1 is related to Green Seal Standard for Commercial Adhesives GS-36 requirements with an effective date of 10/19/2000.

177. Which of the following are true with regard to IEQc4.5: Low-Emitting Materials: Systems Furniture and Seating? (Choose 2)
   a. A project team can use furniture that is certified by GREENGUARD™.
   b. A project team can use furniture that is certified by GREEN-E™.
   c. A project team can limit off-gassing by complying with performance-based standards.
   d. Off-gassing is not related to this credit.

178. Which of the following are true with regard to SSc1, Ch2F6: Light Pollution Reduction? (Choose 2)
   a. A project team can achieve 1 point for SSc1, Ch2F6.
   b. A project team can achieve 2 points for SSc1, Ch2F6.
   c. LZ1-Dark Zone refers to developed areas within rural areas, national parks, and state parks forest land. The population density is less than 100 people per square mile.
   d. LZ1-Dark Zone refers to developed areas within rural areas, national parks, and state parks forest land. The population density is less than 200 people per square mile.

179. Which of the following are true with regard to SSc1, Ch2F6: Light Pollution Reduction? (Choose 2)
   a. It is related to EAc1.1: Optimize Energy Performance: Lighting Power.
   b. You should design building exterior lights and site lights to produce an initial illuminance value no higher than 0.01 vertical and horizontal foot-candles at the site boundary and beyond.
   c. It is related to IEQc6.1: Controllability of Systems: Lighting.
   d. You should document that 5% or less of the total initially designed fixture lumens are emitted at an angle of 90 degree or higher from nadir (straight down).

180. Which of the following are not true with regard to SSc1, Ch2F6: Light Pollution Reduction ? (Choose 2)
    a. A project team needs to submit drawings showing the locations of automatic controls, sequence of operation, and specifications or the building operation plan.
    b. A project team needs to submit drawings and specifications for automatic shading devices, or cutsheets (product data) showing that they block 75% of the light, sequence of operation, and specifications or the building operation plan
    c. Its primary objective is to reward base buildings which reduce development impacts on nocturnal environments, reduce glare, improve nighttime visibility, and reduce sky-glow and light trespass from the site and building to increase night sky access.
    d. It is related to ASTM.

181. Advanced Buildings™ Core Performance™ Guide by New Buildings Institute is related to which of the following?
    a. EAp1: Fundamental Commissioning of the Building Energy Systems
    b. EAc1: Optimize Energy Performance
    c. EAc2: Enhanced Commissioning
    d. None of above

182. ASHRAE 52.2-1999 is related to which of the following? (Choose 2)
    a. EAc1.3: Optimize Energy Performance: HVAC
    b. IEQc3.1: Construction IAQ Management Plan: During Construction
    c. IEQc4.1: Low-Emitting Materials: Adhesive and Sealants
    d. IEQc5: Indoor Chemical and Pollutant Source Control

183. ASHRAE Standard 55-2004 is related to which of the following? (Choose 2)
    a. IEQp1: Minimum IAQ Performance
    b. IEQc3: Construction IAQ Management Plan
    c. IEQc6: Controllability of Systems
    d. IEQc7: Thermal Comfort

184. ASHRAE 62.1-2007 is related to which of the following? (Choose 2)
    a. IEQp1: Minimum IAQ Performance
    b. IEQc1: Outdoor Air Delivery Monitoring
    c. IEQc3: Construction IAQ Management Plan
    d. EAp2: Minimum Energy Performance

185. ASHRAE/IESNA Standard 90.1-2007 is related to which of the following? (Choose 2)
    a. IEQp1: Minimum IAQ Performance
    b. IEQc1: Outdoor Air Delivery Monitoring
    c. EAc1: Optimize Energy Performance
    d. SSc1, Ch2F11: On-Site Renewable Energy

186. ANSI is related to which of the following? (Choose 2)
    a. EAp1: Fundamental Commissioning of the Building Energy Systems
    b. EAp2: Minimum Energy Performance
    c. EAc2: Enhanced Commissioning
    d. IEQp1: Minimum IAQ Performance

187. ASTM is not related to which of the following? (Choose 2)
   a. SSc1, Ch2F1: Brownfield Redevelopment
   b. EAp2: Minimum Energy Performance
   c. IEQp2: Environmental Tobacco Smoke (ETS) Control
   d. IEQc1: Outdoor Air Delivery Monitoring

188. BMP is not related to which of the following? (Choose 3)
   a. SSc1, Ch2F1: Brownfield Redevelopment
   b. SSc6.1: Storm Water Design: Quantity Control
   c. SSc6.2: Storm Water Design: Quality Control
   d. SSc8: Light Pollution Reduction

189. DHS is not related to which of the following? (Choose 1)
   a. IEQc4.1: Low-Emitting Materials: Adhesive and Sealants
   b. IEQc4.2: Low-Emitting Materials: Paints and Coatings
   c. IEQc4.3: Low-Emitting Materials: Carpet Systems
   d. IEQc4.4: Low-Emitting Materials: Composite Wood and Agrifiber Products
   e. IEQc4.5: Low-Emitting Materials: Systems Furniture and Seating

190. California Energy Code Title 24 (T-24) is not related to which of the following? (Choose 3)
   a. SSc1, Ch2F4: Heat Island Effect: Non-Roof
   b. SSc1, Ch2F6: Light Pollution Reduction
   c. EAp1: Fundamental Commissioning of the Building Energy Systems
   d. EAp2: Minimum Energy Performance
   e. EAc1.1: Optimize Energy Performance: Lighting Power

191. CRI is not related to which of the following? (Choose 3)
   a. IEQc4.1: Low-Emitting Materials: Adhesive and Sealants
   b. IEQc4.2: Low-Emitting Materials: Paints and Coatings
   c. IEQc4.3: Low-Emitting Materials: Carpet Systems
   d. IEQc4.4: Low-Emitting Materials: Composite Wood and Agrifiber Products

192. California Energy Code Title 24 (T-24) is related to which of the following? (Choose 2)
   a. EAp1: Fundamental Commissioning of the Building Energy Systems
   b. EAp2: Minimum Energy Performance
   c. EAc1.1: Optimize Energy Performance: Lighting Power
   d. IEQc8.1: Daylight and Views: Daylight

193. Chartered Institution of Building Service Engineers (CIBSE) is related to which of the following?
   a. EAp1: Fundamental Commissioning of the Building Energy Systems
   b. EAp2: Minimum Energy Performance
   c. EAc1: Optimize Energy Performance
   d. IEQc2: Increased Ventilation

194. The Clean Water Act is related to which of the following?
    a. SSc1, Ch2F1: Brownfield Redevelopment
    b. EAp1: Fundamental Commissioning of the Building Energy Systems
    c. IEQc1: Outdoor Air Delivery Monitoring
    d. IEQc2: Increased Ventilation
    e. None of the above

195. Carbon Trust "Good Practice Guide 237" is related to which of the following?
    a. SSc1, Ch2F1: Brownfield Redevelopment
    b. EAp1: Fundamental Commissioning of the Building Energy Systems
    c. IEQc1: Outdoor Air Delivery Monitoring
    d. IEQc2: Increased Ventilation
    e. None of the above

196. Energy Policy Act is related to which of the following? (Choose 2)
    a. WEp1: Water Use Reduction
    b. WEc1: Water Use Reduction
    c. EAp1: Fundamental Commissioning of the Building Energy Systems
    d. IEQc2: Increased Ventilation

197. Energy Star Program, Target Finder Rating Tool is related to which of the following?
    a. SSc1: Site Selection
    b. EAp1: Fundamental Commissioning of the Building Energy Systems
    c. EAp2: Minimum Energy Performance
    d. IEQc1: Outdoor Air Delivery Monitoring

198. Environmental Protection Agency (EPA) is related to which of the following? (Choose 2)
    a. SSc1, Ch2F9: Innovative Wastewater Technologies
    b. WEp1: Water Use Reduction
    c. EAp1: Fundamental Commissioning of the Building Energy Systems
    d. EAp3: Fundamental Refrigerant Management

199. Which of the following normally happen at Construction Documents (CD) Phase of the project? (Choose 2)
    a. MRc1.2: Building Reuse
    b. MRc2: Construction Waste Management
    c. MRc3.1: Materials Reuse
    d. MRc4: Recycled Content

200. Which of the following are true with regard to SSc1, Ch2F6: Light Pollution Reduction ? (Choose 2)
    a. For LZ1, LZ2 and LZ3, if your project boundary is next to a public right-of-way, you can use the curb line instead of the site boundary for light trespass requirements.
    b. For LZ2, LZ3 and LZ4, if your project boundary is next to a public right-of-way, you can use the curb line instead of the site boundary for light trespass requirements.
    c. A project team can use the centerline of the public roadway as the site boundary.
    d. A project team cannot use the centerline of the public roadway as the site boundary.

IV. Answers for the AP ID+C Mock Exam

If you answer 85 of the 100 questions correctly for a section, you have passed the mock exam for that section.

Part One:

1. Answer: d. See LEED checklist. See appendix for link to download the checklists from USGBC website.
2. Answer: b
3. Answer: d
4. Answer: c
5. Answer: a
6. Answer: b
7. Answer: d
8. Answer: a
9. Answer: c
10. Answer: d
11. Answer: d
12. Answer: a and c
13. Answer: c and d
14. Answer: c
15. Answer: d
16. Answer: d
17. Answer: e
18. Answer: a
19. Answer: a. A job captain is a very junior member of a project team; you can<u>not</u> get any LEED point if the LEED AP is not a <u>principal</u> of your team.
20. Answer: c
21. Answer: a and b
22. Answer: b and c. This question gives you a lot of numbers to distract you. All you need to know is that the tenant space (3,8000 s.f.) is less than 75% of the total building area (50,000 s.f.), so d is wrong. a is another obvious wrong answer.
23. Answer: d. The key for this question is the project size: 45,000 gross s.f. < 50,000 gross s.f., so the CxA can be a qualified person on the construction or design team with the necessary experience.
24. Answer: c.
25. Answer: a
26. Answer: b
27. Answer: d
28. Answer: c
29. Answer: b
30. Answer: d
31. Answer: b
32. Answer: a
33. Answer: c
34. Answer: a
35. Answer: c and d
36. Answer: c. See MRc3.1, Materials Reuse
37. Answer: b

38. Answer: a
39. Answer: b
40. Answer: b
41. Answer: b
42. Answer: a and b
43. Answer: a and b
44. Answer: b and c
45. Answer: c
46. Answer: d
47. Answer: a
48. Answer: a, c, and d
49. Answer: c
50. Answer: a, b, and d
51. Answer: b. See IEQc4.2, Low Emitting Materials
52. Answer: a. See IEQp1, Minimum IAQ Performance
53. Answer: b
54. Answer: c
55. Answer: d
56. Answer: c
57. Answer: a
58. Answer: b
59. Answer: b
60. Answer: a, b and c
61. Answer: b and c
62. Answer: d
63. Answer: c
64. Answer: a and d. See SS Credit 3.2.
65. Answer: b, e and f
66. Answer: a
67. Answer: d
68. Answer: b and d
69. Answer: d. See SSc1, Ch2F2, it should be 1 ½ year 24-hour rate.
70. Answer: b
71. Answer: d
72. Answer: c
73. Answer: a
74. Answer: d
75. Answer: a, b and d
76. Answer: c
77. Answer: a
78. Answer: b
79. Answer: d
80. Answer: d
81. Answer: a, b and e
82. Answer: a, c and d
83. Answer: b, c and d
84. Answer: a
85. Answer: a, c and d
86. Answer: a and c

87. Answer: c
88. Answer: a, b, c and e
89. Answer: e
90. Answer: d
91. Answer: a, c and d. WEc1: Water Use Reduction is for INTERIOR water.
92. Answer: a, b and c
93. Answer: d
94. Answer: b and d
95. Answer: a and c
96. Answer: b, c, and d
97. Answer: a
98. Answer: b, c, and e
99. Answer: a and d. b is incorrect because "recycled" in this question refers to "waste recycled" rather than the "recycled content of materials used."
100. Answer: b

Part Two:

101. Answer: a, c, and d
102. Answer: a, c, and d
103. Answer: a, b, and e
104. Answer: a and e. See SSc3.3. You need to provide preferred parking for carpools or vanpools capable of serving 5% or more of tenant occupants (NOT 5% of the total number of parking spaces). Since the question did not give you information on what the occupancy of the tenant space is (grocery store, retail, or office, etc), you cannot determine the number of occupants, and therefore, you cannot determine the exact number of preferred parking needed for carpools or vanpools.
105. Answer: b. The owner or developer must occupy **more than 50%** of the building's **leasable area** to be able to seek LEED-NC certification. Otherwise, the owner or developer should seek LEED-CS certification.
106. Answer: d. MPRs are LEED's **Minimum Program Requirements**. A project must meet MPRs to qualify for LEED certification. One of the MPRs is that the building to site ratio must be 2% or higher. This project's building to site ratio is 1.5%, and therefore does not qualify for LEED certification.
107. Answer: d. Only the project administrator has access to the CIR database and LEED Online after registration. As part of the registration, the LEED project team will have to agree to report post-occupancy water and energy use. No credit will be awarded at the design submittal.
108. Answer: a
109. Answer: a, c and d
110. Answer: c and d. A CIR only provides feedback, and it will not guarantee a credit. After a CIR, you will know that you probably will get a credit but NOT for sure.
111. Answer: b and c. When submitting a CIR, do not mention the contact information, name of the credit, or confidential information, and do not include any attachments.
112. Answer: a and c. The separation of the tasks for the USGBC and the GBCI meets the protocols of the American National Standard Institute (ANSI) and International Organization for Standardization (ISO). A building must be 1,000 s.f. minimum for LEED-NC, LEED Schools, LEED-CS, and LEED-EB: O&M. A LEED project must be a permanent and complete space or building. A LEED project must comply with environmental laws.
113. Answer: a and c. A building must be 250 s.f. minimum for LEED-CI.

114. Answer: c and d. See "**Key stake holders and an integrative approach"** in Chapter 4.
115. Answer: a, c and e.
116. Answer: c, d and e
117. Answer: b and c
118. Answer: a, c and d
119. Answer: c
120. Answer: a
121. Answer: b
122. Answer: b and d
123. Answer: b and d
124. Answer: d
125. Answer: a, d and e
126. Answer: b and d. You can only count a school's bus system as <u>one</u> bus line.
127. Answer: a, b and d
128. Answer: b, c and d.
129. Answer: b and c
130. Answer: b and d.
131. Answer: c and d. "may" is the key word for this question. Items a to b are ALL required (NOT optional) for this credit.
132. Answer: b and d
133. Answer: a and d
134. Answer: a and b
135. Answer: b. The points the project team can get: SS+WE+EA+MR+IEQ = 48, plus a maximum of <u>**6**</u> possible <u>extra</u> points for Innovation in Design, and a maximum of <u>**4**</u> possible <u>extra</u> points for Regional Priority (RP). So, this project can get a maximum of 58 points, or Silver for the building LEED certification. **Note:** There are 6 possible RP points, but you can only pick and choose a maximum of 4 points for each project.
136. Answer: a
137. Answer: b, c and d
138. Answer: a
139. Answer: a and d
140. Answer: c and d
141. Answer: b and c. We are looking for the INCORRECT statements here.
142. Answer: c and e
143. Answer: a, b, and d
144. Answer: c
145. Answer: a and c
146. Answer: a and c
147. Answer: b and d
148. Answer: b and c
149. Answer: c
150. Answer: a
151. Answer: b and d. Make sure you include the ID bonus point in addition to the regular point(s) for MRC5 and MRc6.
152. Answer: d
153. Answer: c and d
154. Answer: a and c
155. Answer: c. This credit is based on cost, and there is more than one kind of recycled content involved, so you can NOT determine the percentage of the cost by weight.

156. Answer: a. Based on cost, the sum of post-consumer recycled content plus ½ of the pre-consumer recycled content = 3 + 5 × ½ = 5.5 million dollars, 5.5 million dollars ÷ 30 million dollars = 18.33 % < 20%. So, the contractor can achieve 1 point for this credit.
157. Answer: c
158. Answer: b
159. Answer: d
160. Answer: c. 5 millions ÷ 30 million dollars =16.67%>10%, so this project can get 1 basic point + 1 bonus point = 2 points for this credit.
161. Answer: b and c
162. Answer: a
163. Answer: b
164. Answer: a
165. Answer: c and d
166. Answer: b and d
167. Answer: a and d
168. Answer: b, c and d
169. Answer: a, b and c
170. Answer: a
171. Answer: a, c and e
172. Answer: d
173. Answer: b and d
174. Answer: c and d
175. Answer: d
176. Answer: a and d
177. Answer: a and c
178. Answer: a and d
179. Answer: a and c
180. Answer: b and d
181. Answer: b
182. Answer: b and d
183. Answer: c and d
184. Answer: a and b
185. Answer: c and d
186. Answer: b and d
187. Answer: b and d
188. Answer: a, b and d
189. Answer: d and e
190. Answer: a, b and c
191. Answer: a, b and d
192. Answer: b and c
193. Answer: d
194. Answer: e
195. Answer: e. See latest USGBC addenda dated 2/1/10 available at USGBC website, it takes precedent over the USGBC reference Guide.
196. Answer: a and b
197. Answer: c
198. Answer: b and d
199. Answer: b and c
200. Answer: b and c

V. How were the LEED AP ID+C mock exams created?

The actual LEED AP ID+C Exam has 200 questions (100 questions for each section) and must be finished within four hours. The raw exam score is converted to a scaled score ranging from 125 to 200. The passing score is 170 or higher.

I tried to be scientific when selecting the mock exam questions, so I based the number of questions for each credit category roughly on the number of points that you can get for that category. The level difficulty for each question was designed to match the official sample questions that can be downloaded from the official GBCI website. Feedback from our readers has indicated that this mock exam is relatively easy when compared to the actual LEED AP ID+C Exam.

VI. Latest trend for LEED exams

Recently, some readers have encountered versions of the LEED exams that contain many questions on refrigerants (CFC, HCFC, and HFC). The following advice will help you answer these questions correctly:

For more information, download the free PDF file called "The Treatment by LEED of the Environmental Impact of HVAC Refrigerants" at the following link:

http://www.gbci.org/Files/References/The-Treatment-by-LEED-of-the-Environmental-Impact-of-HVAC-Refrigerants.pdf

This is a VERY important document that you need to become familiar with. Many real LEED exam questions (CFC, HCFC, HFC, etc.) come from this document. Be familiar with this material.

Pay special attention to the table on ODP and GWP on page 3. You do not have to remember the exact value of all ODPs and GWPs, but you do need to know the rough numbers for various groups of refrigerants.

This latest trend regarding refrigerants (CFC, HCFC, and HFC) for LEED Exams has much to do with LEED v3.0 Credit Weighting. EA (including refrigerants) is the biggest winner in LEED v3.0, meaning the category has MORE questions than any other areas for ALL the LEED exams. See Section entitled "How are LEED credits allocated and weighted?" in Chapter Two of this book.

VII. Where can I find the latest and official sample questions for the LEED AP ID+C Exam?

Answer: You can find them, as well as the exam content, from the candidate handbook, at:
http://www.gbci.org/main-nav/professional-credentials/resources/candidate-handbooks.aspx

VIII. LEED AP ID+C Exam registration

1. How to register for the LEED AP ID+C Exam?

Answer: Per the GBCI, you must create an Eligibility ID at www.GBCI.org. Select the "My Credentials", and then select "Register for an exam" from your "Current Options," and you will find detailed instructions on how to set up an exam time and date with Prometric. You can reschedule or cancel the LEED AP ID+C Exam at www.prometric.com/gbci with your Prometric-issued confirmation number for the exam. You need to bring two forms of ID to the exam site. See www.prometric.com/gbci for a list of exam sites. Call 1-800-795-1747 (within the US) or 202-742-3792 (Outside of the US) or e-mail exam@gbci.org if you have any questions.

2. Important Note: You can download the "LEED AP ID+C Candidate Handbook" from the GBCI website and get all the latest details and procedures. Ideally you should download it and read it carefully at least three weeks before your exam. Refer to the following link:
http://www.gbci.org/main-nav/professional-credentials/resources/candidate-handbooks.aspx

Chapter 7

Frequently Asked Questions (FAQ) and Other Useful Resources

The following are tips on how to pass the LEED exam on the first try with only one week of preparation. I also include my responses to several readers' questions. Hopefully they may help you.

1. I found the reference guide way too tedious. Can I only read your book and just refer to the USGBC reference guide (if one is available for the exam I am taking) when needed?

Response: Yes. That is one way to study.

2. Is one week really enough for me to prepare for the exam while I am working?

Response: Yes, if you can put in 40 to 60 hours during the week, study hard and you can pass the exam. This exam is similar to a history or political science exam; you need to MEMORIZE the information. If you take too long, you will probably forget the information by the time you take the test.

In my book, I give you tips on how to MEMORIZE the information, and I have already highlighted/underlined the most important information that you definitely have to MEMORIZE to pass the exam. It is my goal to use this book to help you to pass the LEED exam with the minimum time and effort. I want to make your life easier.

3. Would you say that if I buy your LEED Exam Guide Series books, I could pass the exam using no other study materials? The books sold on the USGBC website run in the hundreds of dollars, so I would be quite happy if I could buy your book and just use that.

Response: First of all, there are readers who have passed the LEED Exam by reading <u>only</u> my books in the LEED Exam Guides Series (www.ArchiteG.com). My goal is to write one book for each of the LEED exams, and make each of my books stand alone to prepare people for one specific LEED exam.

Secondly, people learn in many different ways. That is why I have added some new advice below for people who learn better by doing practice tests.

If you do the following things, you have a very good chance of passing the LEED exam (NOT a guarantee, <u>nobody</u> can guarantee you will pass):

a. If you study, understand and MEMORIZE all of the information in my book, and do NOT panic when you run into problems you are not familiar with, and use the guess strategy in my book, then you have a very good chance of passing the exam.

You need to UNDERSTAND and MEMORIZE the information in the book and score almost a perfect score on the mock exam in this book. This book will give you the BULK of the most CURRENT information that you need for the specific LEED exam you are taking. You HAVE to know the information in my book in order to pass the exam.

b. If you have not done any LEED projects before, I suggest you also go to the USGBC website and download the latest LEED credit templates for the LEED rating system related to the LEED exam you are taking. Read the templates and become familiar with them. This is important. Refer to the following link:

http://www.usgbc.org/DisplayPage.aspx?CMSPageID=222

c. If you want to be safe and do additional sample tests to find out if you are ready for the real exam, we have other books on various LEED mock exams. Check them out at: www.**GreenExamEducation**.com

The LEED exam is NOT an easy exam, but anyone with a 7th grade education should be able to study and pass the LEED exam if he prepares correctly.

If you have extra time and money, the other books I would recommend are *LEED GA Mock Exams* and the USGBC reference guide, the official book for the LEED exam. I know some people who did not even read the reference guide from cover to cover when they took the exam. They just studied the information in my book, and only referred to the reference guide to look up a few things, and they passed on the first try. Some of my readers have even passed WITHOUT reading the USGBC reference guide AT ALL.

4. I am preparing for the LEED exam. Do I need to read the 2" thick reference?

Response: See answer above.

5. For LEED v3.0, will the total number of points be more than 110 in total if a project gets all of the extra credits and all of the standard credits?

Response: No. For LEED v3.0, there are 100 base points and 10 possible bonus points. There are many ways to get bonus points (extra credits or exemplary performance), but you can have a maximum number of 6 ID bonus points and 4 Regional Priority bonus points. So, the maximum points for ANY project will be 110.

On another note, the older versions of LEED rating systems all have less than 110 possible points except LEED for **Home**, which has 136 possible points.

6. For the exam, do I need to know the project phase in which a specific prerequisite/credit takes place? (i.e., pre-design, schematic design, etc.)

Response: The information on the project phase (NOT LEED submittal phase) for each prerequisite/credit is NOT mentioned in the USGBC reference guide, but it is covered in the USGBC workshops. If it is important enough for the USGBC workshops to cover, then it may show up on the actual LEED exam.

Most, if not all, other third party books completely miss this important information. I cover it for each prerequisite/credit in my book for the LEED exam because I think it is very important.

Some people THINK that the LEED exam ONLY tests information covered by the USGBC reference guide. They are wrong.

The LEED exam does test information NOT covered by the USGBC reference guide at all. This may include the process of LEED submittal and project team coordination, etc.

I would MEMORIZE this information if I were you, because it may show up on the LEED exam. Besides, this information is not hard to memorize once you understand it, and you need to know it to do actual LEED submittal work anyway.

7. Are you writing new versions of books for the new LEED exams? What new books are you writing?

Response: Yes, I am working on other books in the LEED Exam Guide Series. I will be writing one book for each of the LEED exam. See ArchiteG.com for more information.

8. Important documents that you need to download for <u>free</u>, become familiar with and <u>memorize</u>:

Note: GBCI and USGBC change the links to their document every now and then, so, by the time you read this book, they may have changed some of the following links. You can simply go to their main website, search for the document with its name, and should be able to find the most current link. You can use the same technique to search for documents by other organizations.

The main website for the GBCI is:
http://www.gbci.org/

The main website for the USGBC is:
http://www.usgbc.org/

Every LEED exam **always tests** Credit Interpretation Request (CIR). Download the guidelines for CIR customers, read and <u>memorize</u> it:
http://www.gbci.org/Certification/Resources/cirs.aspx
Every LEED exam **always tests** project team coordination. Download *Sustainable Building Technical Manual: Part II,* by Anthony Bernheim and William Reed (1996), read and <u>memorize</u> it:
http://www.gbci.org/Files/References/Sustainable-Building-Technical-Manual-Part-II.pdf

Project registration application and LEED certification process:
http://www.usgbc.org/DisplayPage.aspx?CMSPageID=1497

LEED Online:
https://www.gbci.org/DisplayPage.aspx?CMSPageID=137

9. Important documents that you need to download for <u>free</u>, and become <u>familiar</u> with:

LEED for Operations and Maintenance Reference Guide-Introduction (U.S. Green Building Council, 2008)
https://www.usgbc.org/ShowFile.aspx?DocumentID=4512

LEED for Operations and Maintenance Reference Guide-Glossary (U.S. Green Building Council, 2008) http://www.gbci.org/Files/References/LEED-for-Operations-and-Maintenance-Reference-Guide-Glossary.pdf

LEED for Homes Rating System (U.S. Green Building Council, 2008)
http://www.gbci.org/Files/References/LEED-for-Homes-Rating-System.pdf

Pay special attention to the list of **abbreviations and acronyms** on pages 105–106 and a helpful **glossary of terms** on pages 107–114.

Cost of Green Revisited, by Davis Langdon (2007)
http://www.gbci.org/Files/References/Cost-of-Green-Revisited.pdf

*The Treatment by LEED® of the Environmental Impact of HVAC Refrigerants (*LEED Technical and Scientific Advisory Committee, 2004)
http://www.gbci.org/Files/References/The-Treatment-by-LEED-of-the-Environmental-Impact-of-HVAC-Refrigerants.pdf

Guidance on Innovation and Design (ID) Credits (US Green Building Council, 2004)
http://www.gbci.org/Files/References/Guidance-on-Innovation-and-Design-Credits.pdf

10. Do I need to take many practice questions to prepare for a LEED exam?

Response: There is NO absolutely correct answer to this question. People learn in many different ways. Personally, I am NOT crazy about doing many practice questions. Consider if you do 700 practice questions, not only must you read them all, but each question has at least 4 choices. That totals to at least 2,800 choices, which is a great deal of reading. I have seen some third-party materials that have 1,200 practice questions. That will require even MORE time to go over the materials.

I prefer to spend most of my time reading, digesting, and really understanding the fundamental materials, and MEMORIZE them naturally by rereading the materials multiple times. This is because the fundamental materials for ANY exam will NOT change, and the scope of the exam will NOT change for the same main version of the test (until the exam moves to the next advanced version). However, there are many ways to ask you questions.

If you have a limited amount of time for preparation, it is more efficient for you to focus on the fundamental materials and actually master the knowledge that GBCI wants you to learn. If you can do that, then no matter how GBCI changes the exam format or how GBCI asks the questions, you will do fine in the exam.

Strategy 101 for the LEED AP ID+C Exam is that you must recognize that you have only a limited amount of time to prepare for the exam. Therefore, you must concentrate on the most important contents of the LEED AP ID+C Exam.

The key to passing the LEED AP ID+C Exam, or any other exam, is to know the scope of the exam, and not to read too many books. Select one or two helpful books and focus on them. You must understand the content and memorize it. For your convenience, I have underlined the fundamental information that I think is very important. You definitely need to memorize all the information that I have underlined. You should try to understand the content first, and then memorize the content of the book by rereading it. This is a much better way than "mechanical" memory without understanding.

Most people fail the exam NOT because they are unable to answer the few "advanced" questions on the exam, but because they have read the information but can NOT recall it on the day of the exam. They spend too much time preparing for the exam, drag the preparation process on too long, seek too much information, go to too many Web sites, do too many practice questions and too many mock exams (one or two sets of mock exams are probably sufficient), and spread themselves too thin. They end up missing out on the most important information of the LEED exam, and they will fail.

To me, Memorization and Understanding work hand-in-hand. Understanding always comes first. If you really understand something, then Memorization is easy.

For example, I'll read a book's first chapter very slowly but make sure I <u>really</u> understand everything in it, no matter how long it takes. I do NOT care if others are faster readers than I. Then, I reread the first chapter again. This time, the reading is so much easier, and I can read it much faster. Then I try to retell the contents, focusing on substance, not the format or any particular order of things. This is a very good way for me to understand and digest the material, while <u>absorbing</u> and <u>memorizing</u> the content.

I then repeat the same procedure for each chapter, and then reread the book until I take the exam. This achieves two purposes:

a. I keep reinforcing the important materials that I already have memorized and fight against the human brain's natural tendency to forget things.

b. I also understand the content of the book much better by reading it multiple times.

If I were to attempt to memorize something without understanding it first, it would be very difficult for me to do so. Even if I were to memorize it, I would likely forget it quickly.

Appendixes

1. Default occupancy factors

Occupancy	Gross sf per occupant	
	Transient Occupant	FTE
Educational, Daycare	630	105
Educational, K–12	1,300	140
Educational, Postsecondary	2,100	150
Grocery store	550	115
Hotel	1,500	700
Laboratory or R&D	400	0
Office, Medical	225	330
Office, General	250	0
Retail, General	550	130
Retail or Service (auto, financial, etc.)	600	130
Restaurant	435	95
Warehouse, Distribution	2,500	0
Warehouse, Storage	20,000	0

Note: This table is for projects (like CS) where the final occupant count is not available. If your project's occupancy factors are not listed above, you can use a comparable building to show the average gross sf per occupant for your building's use.

2. Important resources and further study materials you can download for free

Energy Performance of LEED® for New Construction Buildings: Final Report, by Cathy Turner and Mark Frankel (2008):
http://www.gbci.org/Files/References/Energy-Performance-of-LEED-for-New-Construction-Buildings-Final-Report.pdf

Foundations of the Leadership in Energy and Environmental Design Environmental Rating System: A Tool for Market Transformation (LEED Steering Committee, 2006):
http://www.gbci.org/Files/References/Foundations-of-the-Leadership-in-Energy-and-Environmental-Design-Environmental-Rating-System-A-Tool-for-Market-Transformation.pdf

AIA Integrated Project Delivery: A Guide (www.aia.org):
http://www.aia.org/contractdocs/AIAS077630

Review of ANSI/ASHRAE Standard 62.1-2007: Ventilation for Acceptable Indoor Air Quality, by Brian Kareis:
http://www.workplace-hygiene.com/articles/ANSI-ASHRAE-3.html

Best Practices of ISO-14021: Self-Declared Environmental Claims, by Kun-Mo Lee and Haruo Uehara (2003):

http://www.ecodesign-company.com/documents/BestPracticeISO14021.pdf

Bureau of Labor Statistics (www.bls.gov)

International Code Council (www.iccsafe.org)

Americans with Disabilities Act (ADA): Standards for Accessible Design (www.ada.gov):
http://www.ada.gov/stdspdf.htm

GSA 2003 Facilities Standards (General Services Administration, 2003):
http://www.gbci.org/Files/References/GSA-2003-facilities-standards.pdf

Guide to Purchasing Green Power (Environmental Protection Agency, 2004):
http://www.gbci.org/Files/References/Guide-to-Purchasing-Green-Power.pdf

USGBC Definitions:
https://www.usgbc.org/ShowFile.aspx?DocumentID=5744

3. Annotated bibliography

Chen, Gang. ***LEED Green Associate Exam Guide (LEED GA):*** *Comprehensive Study Materials, Sample Questions, Mock Exam, Green Building LEED Certification, and Sustainability (LEED v3.0)*, Book 2, LEED Exam Guide series, ArchiteG.com (3rd Printing of the 3rd Edition). ArchiteG, Inc. Published December 21, 2011. This is a very comprehensive and concise book on the LEED Green Associate Exam. Some readers have passed the LEED Green Associate Exam by studying this book for 3 days.

Chen, Gang. ***LEED GA MOCK EXAMS:*** *Questions, Answers, and Explanations: A Must-Have for the LEED Green Associate Exam, Green Building LEED Certification, and Sustainability*. ArchiteG, Inc, 2010. This is a companion to LEED Green Associate Exam Guide. It includes 200 questions, answers, and explanation, and is very close to the real LEED Green Associate Exam.

4. Valuable Web sites and links

a. The Official Web sites for the U.S. Green Building Council (USGBC):
http://www.usgbc.org/
http://www.Greenbuild365.org

Pay special attention to the purpose of LEED Online, LEED project registration, LEED certification content, LEED reference guide introductions, LEED rating systems, and checklists.

You can download or purchase the following useful documents from the USGBC or GBCI Web site:
Latest and official LEED exam candidate handbooks including an exam content outline and sample questions:
http://www.gbci.org/main-nav/professional-credentials/resources/candidate-handbooks.aspx

LEED Reference Guides, **LEED Rating System Selection Policy**, and various versions of LEED Green Building Rating Systems and Project Checklist:
http://www.usgbc.org/projecttools

Read the **LEED Rating System Selection Policy** at least three times, because it is VERY important, and it tells you which LEED system to use.

LEED 2009 Vision and Executive Summary:
http://www.usgbc.org/ShowFile.aspx?DocumentID=4121

USGBC issue LEED Addenda for various LEED Green Building Rating **Systems** and **reference guides** on a quarterly basis. **Make sure you download the latest LEED Addenda** related to your exam and read them at least three times. See link below for detailed information:
http://www.usgbc.org/addenda

b. Natural Resources Defense Council:
http://www.nrdc.org/

c. Environmental Construction + Design - Green Book (Offers print magazine and online environmental products and services resources guide):
http://www.edcmag.com/greenbook

d. Cool Roof Rating Council Web site:
http://www.coolroofs.org

5. Important Items Covered by the Second Edition of *Green Building and LEED Core Concepts Guide*

Starting on December 1, 2011, GBCI began to draw LEED AP ID+C Exam questions from the second edition of *Green Building and LEED Core Concepts Guide*. The following are some "new" and important items covered by this edition:

adaptive reuse: Designing and constructing a building to accommodate a future use that is different from its original use.

biomimicry: Learning from nature and designing systems using principles that have been tested in nature for millions of years.

carbon overlay: LEED credit weighting based on each credit's impact on reducing carbon footprint.

charrettes: Intensive (design) workshops.

cradle to cradle: A method where materials are used in a closed system and generate no waste.

cradle to grave: A process that examines materials from their point of extraction to disposal.

closed system: There is no "away." Everything goes somewhere within the system, the waste generated by a process becomes the "food" of another process. Nature is a closed system.

embodied energy: The total energy consumed by extracting, harvesting, manufacturing, transporting, installing, and using a material through its entire life cycle.

ENERGY STAR's Portfolio Manager: An online management tool for tracking and evaluating water and energy use. An ENERGY STAR Portfolio Manager score of 50 means a building is at national average energy use level for its category. A score higher than 50 means a building is more energy efficient than the national average energy use level for its category. The higher the score, the better.

evapotranspiration: Loss of water due to evaporation.

externalities: Benefits or costs that are NOT part of a transaction.

feedback loop: Information flows within a system that allows the system to adjust itself. A thermostat or melting snow is an example of negative feedback loop. Population growth, heat island effect, or climate change is a positive feedback loop. Positive feedback loop can create chaos in a system.

International Green Construction Code (IGCC): A national model green building code published by International Code Council (ICC).

integrated process: Emphasizes communications and interactions among stakeholders throughout the life of a project. Integrated process is a holistic decision making process based on systems thinking and life cycle approach.

interative process: A repetitive and circular process that helps a team to define goals and check ideas against these goals.

Integrated Pest Management (IPM): A sustainable approach to pest management.

LEED interpretations: Precedent-setting (project credit interpretation) rulings. A project team can opt into the LEED interpretation process when submitting an inquiry to GBCI.

leverage points: Places where a small interventions can generate big changes.

life cycle approach: Looking at a product or building through its entire life cycle.

life cycle assessment (LCA): Use life cycle thinking in environmental issues.

life cycle costing: Looking at the cost of purchasing and operating a building or product, and the relative savings.

low impact development (LID): A land development approach mimicking natural systems and managing storm water as close to the source as possible.

"Net-Zero": A project, which doesn't use any more resources than what it can produce. Similar concepts include carbon neutrality and water balance.

negative feedback loop: A signal for the system to stop changes when a response is not needed anymore.

open system: Resources are brought from the outside, consumed, and then disposed of as waste to the outside.

permaculture: Designing human habitats and agriculture systems based on models and relationships found in nature.

positive feedback loop: A stimulus causes an effect and encourages the loop to produce more of this effect.

Prius effect: Provides real time feedback of energy use so that users can adjust behaviors to save energy.

Project CIRs: LEED credit interpretation rulings for specific project circumstances.

retrocommissioning: A building tune-up that restores efficiency and improves performance.

regenerative: Regenerative buildings and communities evolve with living systems and help to renew resources and life. Regenerative projects generate electricity and sell the excess back to the grid, as well as return water to nature, which is cleaner than it was before use.

systems thinking: In a system, each component affects many other components. They are all related to each other.

Wingspread Principles on the US Response to Global Warming: A set of principles signed by organizations and individuals to express their commitment to address global warming. It calls for 60% to 80% reduction of green house gas emission by midcentury (based on 1990 levels).

Back page promotion

You may be interested in some other books written by Gang Chen:

A. **ARE Mock Exam series. See following link:**
   http://www.GreenExamEducation.com

B. **LEED Exam Guides series. See following link:**
   http://www.GreenExamEducation.com

C. ***Building Construction:*** *Project Management, Construction Administration, Drawings, Specs, Detailing Tips, Schedules, Checklists, and Secrets Others Don't Tell You (Architectural Practice Simplified, 2nd edition)*
   http://www.ArchiteG.com

D. ***Planting Design Illustrated***
   http://outskirtspress.com/agent.php?key=11011&page=GangChen

ARE Mock Exam Series

Published ARE books:

Construction Documents and Service (CDS) Are Mock Exam (Architect Registration Exam): ARE Overview, Exam Prep Tips, Multiple-Choice Questions and Graphic Vignettes, Solutions and Explanations
ISBN-13: 9781612650005
(Published May 22, 2011)

Building Design and Construction Systems (BDCS) ARE Mock Exam (Architect Registration Exam): ARE Overview, Exam Prep Tips, Multiple-Choice Questions and Graphic Vignettes, Solutions and Explanations
ISBN-13: 9781612650029
(Published July 12, 2011)

Building Systems (BS) ARE Mock Exam (Architect Registration Exam): ARE Overview, Exam Prep Tips, Multiple-Choice Questions and Graphic Vignettes, Solutions and Explanations
ISBN-13: 9781612650036
(Published October 28, 2011)

Schematic Design (SD) ARE Mock Exam (Architect Registration Exam): ARE Overview, Exam Prep Tips, Graphic Vignettes, Solutions and Explanations
ISBN: 9781612650050
(Published November 18, 2011)

Upcoming ARE books:

Other books in the ARE Mock Exam Series are being produced. Our goal is to produce one mock exam book PLUS one guidebook for each of the ARE exam division.

See the following link for the latest information:
http://www.GreenExamEducation.com

LEED Exam Guides Series*: Comprehensive Study Materials, Sample Questions, Mock Exam, Building LEED Certification and Going Green*

LEED (Leadership in Energy and Environmental Design) is the most important trend of development, and it is revolutionizing the construction industry. It has gained tremendous momentum and has a profound impact on our environment.

From LEED Exam Guides series, you will learn how to

1. Pass the LEED Green Associate Exam and various LEED AP + exams (each book will help you with a specific LEED exam).

2. Register and certify a building for LEED certification.

3. Understand the intent for each LEED prerequisite and credit.

4. Calculate points for a LEED credit.

5. Identify the responsible party for each prerequisite and credit.

6. Earn extra credit (exemplary performance) for LEED.

7. Implement the local codes and building standards for prerequisites and credit.

8. Receive points for categories not yet clearly defined by USGBC.

There is currently NO official book on the LEED Green Associate Exam, and most of the existing books on LEED and LEED AP are too expensive and too complicated to be practical and helpful. The pocket guides in LEED Exam Guides series fill in the blanks, demystify LEED, and uncover the tips, codes, and jargon for LEED as well as the true meaning of "going green." They will set up a solid foundation and fundamental framework of LEED for you. Each book in the LEED Exam Guides series covers every aspect of one or more specific LEED rating system(s) in plain and concise language and makes this information understandable to all people.

These pocket guides are small and easy to carry around. You can read them whenever you have a few extra minutes. They are indispensable books for all people—administrators; developers; contractors; architects; landscape architects; civil, mechanical, electrical, and plumbing engineers; interns; drafters; designers; and other design professionals.

Why is the LEED Exam Guides series needed?

A number of books are available that you can use to prepare for the LEED Exams:

1. *USGBC Reference Guides*. You need to select the correct version of the *Reference Guide* for your exam.

   The *USGBC Reference Guides* are comprehensive, but they give too much information. For example, *The LEED 2009 Reference Guide for Green Building Design and Construction (BD&C)* has about 700 oversized pages. Many of the calculations in the books are too detailed for the exam. They are also expensive (approximately $200 each, so most people may not buy them for their personal use, but instead, will seek to share an office copy).

   It is good to read a reference guide from cover to cover if you have the time. The problem is not too many people have time to read the whole reference guide. Even if you do read the whole guide, you may not remember the important issues to pass the LEED exam. You need to reread the material several times before you can remember much of it.

   Reading the reference guide from cover to cover without a guidebook is a difficult and inefficient way of preparing for the LEED AP Exam, because you do NOT know what USGBC and GBCI are looking for in the exam.

2. The USGBC workshops and related handouts are concise, but they do not cover extra credits (exemplary performance). The workshops are expensive, costing approximately $450 each.

3. Various books published by a third party are available online. However, most of them are not very helpful.

   There are many books on LEED, but not all are useful.

   LEED Exam Guides series will fill in the blanks and become a valuable, reliable source:

   a. They will give you more information for your money. Each of the books in the LEED Exam Guides series has more information than the related USGBC workshops.

   b. They are exam-oriented and more effective than the USGBC reference guides.

   c. They are better than most, if not all, of the other third-party books. They give you comprehensive study materials, sample questions and answers, mock exams and answers, and critical information on building LEED certification and going green. Other third-party books only give you a fraction of the information.

   d. They are comprehensive yet concise. They are small and easy to carry around. You can read them whenever you have a few extra minutes.

   e. They are great timesavers. I have highlighted the important information that you need to understand and MEMORIZE. I also make some acronyms and short sentences to help you easily remember the credit names.

It should take you about 1 or 2 weeks of full-time study to pass each of the LEED exams. I have met people who have spent 40 hours to study and passed the exams.

You can find sample texts and other information on the LEED Exam Guides series in customer discussion sections under each of my book's listing.

What others are saying about *LEED Green Associate Exam Guide* (Book 2, LEED Exam Guide series):

"Finally! A comprehensive study tool for LEED Green Associate Exam Prep!

"I took the 1-day Green LEED GA course and walked away with a power point binder printed in very small print—which was missing MUCH of the required information (although I didn't know it at the time). I studied my little heart out and took the test, only to fail it by 1 point. Turns out I did NOT study all the material I needed to in order to pass the test. I found this book, read it, marked it up, retook the test, and passed it with a 95%. Look, we all know the LEED Green Associate Exam is new and the resources for study are VERY limited. This one's the VERY best out there right now. I highly recommend it."
—ConsultantVA

"Complete overview for the LEED Green Associate Exam

"I studied this book for about 3 days and passed the exam ... if you are truly interested in learning about the LEED system and green building design, this is a great place to start."
—K.A. Evans

"A Wonderful Guide for the LEED Green Associate Exam

"After deciding to take the LEED Green Associate exam, I started to look for the best possible study materials and resources. From what I thought would be a relatively easy task, it turned into a tedious endeavor. I realized that there are vast amounts of third-party guides and handbooks. Since the official sites offer little to no help, it became clear to me that my best chance to succeed and pass this exam would be to find the most comprehensive study guide that would not only teach me the topics, but would also give me a great background and understanding of what LEED actually is. Once I stumbled upon Mr. Chen's book, all my needs were answered. This is a great study guide that will give the reader the most complete view of the LEED exam and all that it entails.

"The book is written in an easy-to-understand language and brings up great examples, tying the material to the real world. The information is presented in a coherent and logical way, which optimizes the learning process and does not go into details that will not be needed for the LEED Green Associate Exam, as many other guides do. This book stays dead on topic and keeps the reader interested in the material.

"I highly recommend this book to anyone that is considering the LEED Green Associate Exam. I learned a great deal from this guide, and I am feeling very confident about my chances for passing my upcoming exam."
—Pavel Geystrin

"Easy to read, easy to understand

"I have read through the book once and found it to be the perfect study guide for me. The author does a great job of helping you get into the right frame of mind for the content of the exam. I had started by studying the Green Building Design and Construction reference guide for LEED projects produced by the USGBC. That was the wrong approach, simply too much information with very little retention. At 636 pages in textbook format, it would have been a daunting task to get through it. Gang Chen breaks down the points, helping to minimize the amount of information but maximizing the content I was able to absorb. I plan on going through the book a few more times, and I now believe I have the right information to pass the LEED Green Associate Exam."
—Brian Hochstein

"All in one—LEED GA prep material

"Since the LEED Green Associate exam is a newer addition by USGBC, there is not much information regarding study material for this exam. When I started looking around for material, I got really confused about what material I should buy. This LEED GA guide by Gang Chen is an answer to all my worries! It is a very precise book with lots of information, like how to approach the exam, what to study and what to skip, links to online material, and tips and tricks for passing the exam. It is like the 'one stop shop' for the LEED Green Associate Exam. I think this book can also be a good reference guide for green building professionals. A must-have!"
—SwatiD

"An ESSENTIAL LEED Green Associate Exam Reference Guide

"This book is an invaluable tool in preparation for the LEED Green Associate (GA) Exam. As a practicing professional in the consulting realm, I found this book to be all-inclusive of the preparatory material needed for sitting the exam. The information provides clarity to the fundamental and advanced concepts of what LEED aims to achieve. A tremendous benefit is the connectivity of the concepts with real-world applications.

"The author, Gang Chen, provides a vast amount of knowledge in a very clear, concise, and logical media. For those that have not picked up a textbook in a while, it is very manageable to extract the needed information from this book. If you are taking the exam, do yourself a favor and purchase a copy of this great guide. Applicable fields: Civil Engineering, Architectural Design, MEP, and General Land Development."
—Edwin L. Tamang

Note: Other books in the **LEED Exam Guides series** are in the process of being produced. At least **One book will eventually be produced for each of the LEED exams.** The series include:

LEED Green Associate Exam Guide (LEED GA): *Comprehensive Study Materials, Sample Questions, Mock Exam, Green Building LEED Certification, and Sustainability (LEED v3.0)*, Book 2, LEED Exam Guide series, ArchiteG.com (3rd Printing of the 3rd Edition), Book 2, LEED Exam Guide series, ArchiteG.com (Published December 21, 2011)

LEED GA MOCK EXAMS: *Questions, Answers, and Explanations: A Must-Have for the LEED Green Associate Exam, Green Building LEED Certification, and Sustainability*, LEED Exam Guide series, ArchiteG.com (Published August 6, 2010)

LEED BD&C EXAM GUIDE: *A Must-Have for the LEED AP BD+C Exam: Comprehensive Study Materials, Sample Questions, Mock Exam, Green Building Design and Construction, LEED Certification, and Sustainability* (2nd Edition), LEED Exam Guide series, ArchiteG.com (Published December 26, 2011)

LEED BD&C MOCK EXAMS: *Questions, Answers, and Explanations: A Must-Have for the LEED AP BD+C Exam, Green Building LEED Certification, and Sustainability*, LEED Exam Guide series, ArchiteG.com (Published November 26, 2011)

LEED AP Exam Guide: *Study Materials, Sample Questions, Mock Exam, Building LEED Certification (LEED NC v2.2), and Going Green,* LEED Exam Guides series, LEEDSeries.com (Published on 9/23/2008).

LEED ID&C EXAM GUIDE: *A Must-Have for the LEED AP ID+C Exam: Comprehensive Study Materials, Sample Questions, Mock Exam, Green Interior Design and Construction, LEED Certification, and Sustainability,* LEED Exam Guide series, ArchiteG.com (Published March 8, 2010)

LEED O&M MOCK EXAMS: *Questions, Answers, and Explanations: A Must-Have for the LEED O&M Exam, Green Building LEED Certification, and Sustainability*, LEED Exam Guide series, ArchiteG.com (Published September 28, 2010)

LEED O&M EXAM GUIDE: *A Must-Have for the LEED AP O+M Exam: Comprehensive Study Materials, Sample Questions, Mock Exam, Green Building Operations and Maintenance, LEED Certification, and Sustainability (LEED v3.0)*, LEED Exam Guide series, ArchiteG.com

LEED HOMES EXAM GUIDE: *A Must-Have for the LEED AP Homes Exam: Comprehensive Study Materials, Sample Questions, Mock Exam, Green Building LEED Certification, and Sustainability*, LEED Exam Guide series, ArchiteG.com

LEED ND EXAM GUIDE: *A Must-Have for the LEED AP Neighborhood Development Exam: Comprehensive Study Materials, Sample Questions, Mock Exam, Green Building LEED Certification, and Sustainability*, LEED Exam Guide series, ArchiteG.com

How to order these books:
You can order the books listed above at:
http://www.GreenExamEducation.com

OR
http://www.ArchiteG.com

Building Construction

Project Management, Construction Administration, Drawings, Specs, Detailing Tips, Schedules, Checklists, and Secrets Others Don't Tell You (Architectural Practice Simplified, 2nd edition)

Learn the Tips, Become One of Those Who Know Building Construction and Architectural Practice, and Thrive!

For architectural practice and building design and construction industry, there are two kinds of people: those who know, and those who don't. The tips of building design and construction and project management have been undercover—until now.

Most of the existing books on building construction and architectural practice are too expensive, too complicated, and too long to be practical and helpful. This book simplifies the process to make it easier to understand and uncovers the tips of building design and construction and project management. It sets up a solid foundation and fundamental framework for this field. It covers every aspect of building construction and architectural practice in plain and concise language and introduces it to all people. Through practical case studies, it demonstrates the efficient and proper ways to handle various issues and problems in architectural practice and building design and construction industry.

It is for ordinary people and aspiring young architects as well as seasoned professionals in the construction industry. For ordinary people, it uncovers the tips of building construction; for aspiring architects, it works as a construction industry survival guide and a guidebook to shorten the process in mastering architectural practice and climbing up the professional ladder; for seasoned architects, it has many checklists to refresh their memory. It is an indispensable reference book for ordinary people, architectural students, interns, drafters, designers, seasoned architects, engineers, construction administrators, superintendents, construction managers, contractors, and developers.

You will learn:

1. How to develop your business and work with your client.
2. The entire process of building design and construction, including programming, entitlement, schematic design, design development, construction documents, bidding, and construction administration.
3. How to coordinate with governing agencies, including a county's health department and a city's planning, building, fire, public works departments, etc.
4. How to coordinate with your consultants, including soils, civil, structural, electrical, mechanical, plumbing engineers, landscape architects, etc.
5. How to create and use your own checklists to do quality control of your construction documents.
6. How to use various logs (i.e., RFI log, submittal log, field visit log, etc.) and lists (contact list, document control list, distribution list, etc.) to organize and simplify your work.
7. How to respond to RFI, issue CCDs, review change orders, submittals, etc.
8. How to make your architectural practice a profitable and successful business.

Planting Design Illustrated

A Must-Have for Landscape Architecture: A Holistic Garden Design Guide with Architectural and Horticultural Insight, and Ideas from Famous Gardens in Major Civilizations

One of the most significant books on landscaping!

This is one of the most comprehensive books on planting design. It fills in the blanks of the field and introduces poetry, painting, and symbolism into planting design. It covers in detail the two major systems of planting design: formal planting design and naturalistic planting design. It has numerous line drawings and photos to illustrate the planting design concepts and principles. Through in-depth discussions of historical precedents and practical case studies, it uncovers the fundamental design principles and concepts, as well as the underpinning philosophy for planting design. It is an indispensable reference book for landscape architecture students, designers, architects, urban planners, and ordinary garden lovers.

What Others Are Saying About *Planting Design Illustrated* …

"I found this book to be absolutely fascinating. You will need to concentrate while reading it, but the effort will be well worth your time."
—Bobbie Schwartz, former president of APLD (Association of Professional Landscape Designers) and author of *The Design Puzzle: Putting the Pieces Together.*

"This is a book that you have to read, and it is more than well worth your time. Gang Chen takes you well beyond what you will learn in other books about basic principles like color, texture, and mass."
—Jane Berger, editor & publisher of gardendesignonline

"As a longtime consumer of gardening books, I am impressed with Gang Chen's inclusion of new information on planting design theory for Chinese and Japanese gardens. Many gardening books discuss the beauty of Japanese gardens, and a few discuss the unique charms of Chinese gardens, but this one explains how Japanese and Chinese history, as well as geography and artistic traditions, bear on the development of each country's style. The material on traditional Western garden planting is thorough and inspiring, too. *Planting Design Illustrated* definitely rewards repeated reading and study. Any garden designer will read it with profit."
—Jan Whitner, editor of the *Washington Park Arboretum Bulletin*

"Enhanced with an annotated bibliography and informative appendices, *Planting Design Illustrated* offers an especially "reader friendly" and practical guide that makes it a very strongly recommended addition to personal, professional, academic, and community library gardening & landscaping reference collection and supplemental reading list."
—Midwest Book Review

"Where to start? *Planting Design Illustrated* is, above all, fascinating and refreshing! Not something the lay reader encounters every day, the book presents an unlikely topic in an easily digestible, easy-to-follow way. It is superbly organized with a comprehensive table of contents, bibliography, and appendices. The

writing, though expertly informative, maintains its accessibility throughout and is a joy to read. The detailed and beautiful illustrations expanding on the concepts presented were my favorite portion. One of the finest books I've encountered in this contest in the past 5 years."
—Writer's Digest 16th Annual International Self-Published Book Awards Judge's Commentary

"The work in my view has incredible application to planting design generally and a system approach to what is a very difficult subject to teach, at least in my experience. Also featured is a very beautiful philosophy of garden design principles bordering poetry. It's my strong conviction that this work needs to see the light of day by being published for the use of professionals, students & garden enthusiasts."
—Donald C. Brinkerhoff, FASLA, chairman and CEO of Lifescapes International, Inc.

Index

71218785R00138

Made in the USA
Columbia, SC
23 May 2017